Fatigue and Corrosion in Metals

Fatigue and Corrosion in Metals

Editor

Vivek Tiwari

Fatigue and Corrosion in Metals

Edited by **Vivek Tiwari**

Printed in 2017

ISBN: 978-1-68117-217-0

Library of Congress Control Number: 2015936578

© 2016 by
SCITUS Academics LLC,
616, Corporate Way, Suite 2, 4766,
Valley Cottage, NY 10989

www.scitusacademics.com

.

Contents

Preface

Corrosion fatigue is fatigue in a corrosive environment. It is the mechanical degradation of a material under the joint action of corrosion and cyclic loading. Nearly all engineering structures experience some form of alternating stress, and are exposed to harmful environments during their service life. The environment plays a significant role in the fatigue of high-strength structural materials like steel, aluminum alloys and titanium alloys. Materials with high specific strength are being developed to meet the requirements of advancing technology. However, their usefulness depends to a large extent on the extent to which they resist corrosion fatigue. The effects of corrosive environments on the fatigue behavior of metals were studied as early as 1930. The phenomenon should not be confused with stress corrosion cracking, where corrosion (such as pitting) leads to the development of brittle cracks, growth and failure. The only requirement for corrosion fatigue is that the sample be under tensile stress.

Editor

Exfoliation Corrosion and Pitting Corrosion and Their Role in Fatigue Predictive Modeling: State-of-the-Art Review

David W. Hoeppner and Carlos A. Arriscorreta

Department of Mechanical Engineering, University of Utah, 50 South Central Campus Drive, Room 2010, Salt Lake City, UT 84112, USA

ABSTRACT

Intergranular attack (IG) and exfoliation corrosion (EC) have a detrimental impact on the structural integrity of aircraft structures of all types. Understanding the mechanisms and methods for dealing with these processes and with corrosion in general has been and is critical to the safety of critical components of aircraft. Discussion of cases where IG attack and exfoliation caused issues in structural integrity in aircraft in operational fleets is presented herein along with a much more detailed presentation of the issues involved in dealing with corrosion of aircraft. Issues of corrosion and fatigue related to the structural integrity of aging aircraft are introduced herein. Mechanisms of pitting nucleation are discussed which include adsorption-induced, ion migration-penetration, and chemicomechanical film breakdown theories. In addition, pitting corrosion (PC) fatigue models are presented as well as a critical assessment of their application to aircraft structures and materials. Finally environmental effects on short crack behavior of materials are discussed, and a compilation of definitions related to corrosion and fatigue are presented.

INTRODUCTION

This paper deals with the effects of intergranular attack and exfoliation corrosion on structural integrity of aircraft structures and materials with emphasis on aluminum alloys used over many decades for airframe components of military, commercial, and general aviation aircraft. Aluminum alloys have been the material of choice for many components of airframes in the past and remain so even though some aircraft are using more titanium alloys and resin-based composites in many airframe components. The general background on phases of life and methods for dealing with corrosion in general and aspects of HOLSIP (Holistic Structural Integrity Processes) paradigm are presented to some extent. (See http://www.holsip.com/). This is followed by a discussion of corrosion effects on SI (Structural Integrity) with some details provided on significant effects of corrosion on maintainability and reliability of structures with extensive background material. Subsequently a section that describes intergranular attack and exfoliation in general terms follows which then is followed by a discussion of cases where IG attack and exfoliation caused significant structural integrity issues in aircraft in operational fleets. Studies oriented toward evaluating the effects of IG and exfoliation on fatigue behavior with emphasis on the long crack aspects are presented. The final section then presents recommended studies in order to develop and validate models to allow prediction and management of IG attack and exfoliation as part of a Holistic Structural Integrity Processes paradigm [1–67], (numbers in parentheses refer to the references in order of appearance).

Phases of Life and Modeling

The phases of life of a structure may be classified according to the division in the Table 1. Thus, the total life (L_T) of a structure is $L_T=L_1+L_2+L_3+L_4$. Figure 1 presents a depiction of the degradation process from a holistic perspective. The regions shown in Figure 1, for example, 1, 2, 3, and 4, illustrate the portion of life, on the abscissa, and the corresponding growth in discontinuity size plotted schematically on the ordinate. This paper concentrates on the phases of life L_1 and L_2. that is, the corrosion process or processes that results in the formation or nucleation of a specific form of corrosion generating a specific form of discontinuity that is not necessarily a crack-like discontinuity and the development of short cracks and their propagation from the initial discontinuity state or from the evolved or modified discontinuity state formed by the mechanism in question. The requirement of the community to come up with design methods to deal with corrosion or other time-based degradation, that is, fatigue, creep, and wear is essential and some of the elements are depicted in Figure 2. This figure illustrates that most of the quantitative methods that have been developed used the concepts of mechanics of materials with an incorporation of fracture mechanics.

Table 1: Phases of Life. See Figure 1. {From Hoeppner, 1972 [67], 1981 [38], 1985 [39]}

(i) Formation or nucleation of degradation/damage by a specificphysical or corrosion process interacting with the fatigue processif appropriate. Corrosion and other processes may act alone toform/nucleate the damage. A transition from theformation/nucleation stage to the next phase must occur. Phase L1 to some other phase.
(ii) Microstructurally dominated crack linkup and propagation ("short" or "small" crack regime). Phase L2.
(iii) Crack propagation in the regime where LEFM, EPFM, or FPFM may be applied both for analysis and material characterization (the "long" crack regime). Phase L3.
(iv) Final instability. Phase L4.

NOTE: In some cases in practice not all the phases cited above occur.

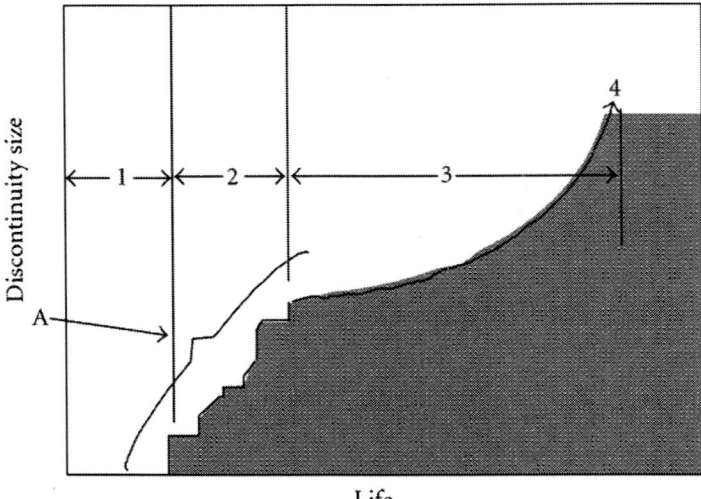

A = "first" detectable crack
1. Nucleation phase, "no crack"
2. "small crack" phase steps related to local
   structure (anisotropy)
3. Stress-dominated crack growth, LEFM, EPFM
4. Crack at length to produce instability

Figure 1: A depiction of the degradation process {after Hoeppner-1972 [67], 1981 [38], 1985 [39]}.

Nucleation	"small crack" growth	Stress-dominated Brack growth	Failure (fracture)
Material failure mechanism with appropriate stress/strain life data	Crack prop. threshold related to structure (micro)	Fracture mechanics • Similitude • Boundary cond. ⟨LEFM EPFM⟩ ?	Klc etc. C.O.D Tensile/ Compressive buckling

Nucleated discontinuity (not inherent) type, size, location	Structure-dominated crack growth	Data base Appropriate stress intensity factor	
Presence of malignant D ,H	Mechanisms, rate	Initial D ,H size, location, type	
Possibility of extraneous effects Corrosion Fretting Creep Mechanical damage	Onset of Stress-dominated crack growth Effects of • R ratio • Stress state • Environment ← chem • Spectrum -waveform	Effects of • R ratio • Stress state • Environment $\leftarrow$ chem • Spectrum -waveform	

Figure 2: Methods for each life phase {after Hoeppner-1972 [67], 1981 [38], 1985 [39]}

The sections of this paper that follow will discuss the following major areas:

- general effects of corrosion on structural integrity;
- intergranular attack and exfoliation corrosion (EC) in aircraft structural aluminum alloys;
- Efforts to date on modeling effects of exfoliation corrosion in aircraft structure with emphasis on fatigue and fatigue crack propagation behavior.

The issue of the effects of corrosion on structural integrity of aircraft has been a question of concern for some time [1–36]. The potential effects are many and they can be categorized as follows.

(An attempt has been made to provide as simple a statement of each potential problem as possible. In the discussion below the use of the terms global and local refers to the likely extent of the corrosion on the surface of a component. Global means the corrosion would be found on much of the component whereas local means the corrosion may be localized to only small, local areas.)

- Reduction of section with a concomitant increase in stress (e.g., thickness change, etc.). Global or local.
- Production of stress concentration. Local.
- Nucleation of cracks. Local, possibly global. Source of Multiple-site cracking.
- Production of corrosion debris. This may result in surface pillowing by various means, which may significantly change the stress state and structural behavior. Local and global.
- Creation of a situation that causes the surfaces to malfunction. Local and global.
- Cause environmentally assisted crack growth (EACG) under cyclic (corrosion fatigue or corrosion-fatigue) or sustained loading (SCC) conditions. Local.
- Create a damage state that is missed in inspection when the inspection plan was not developed for corrosion or when corrosion is missed. Local and global.
- Change the structurally significant item due to the creation of a damage state not envisioned in the structural damage analysis or fatigue and strength analysis. If the SSI is specified, for example, by location of maximum stress or strain, then the corrosion may cause another area(s) to become significant. Local or global.
- Create an embrittlement condition in the material that subsequently affects behavior. Local or global.
- Create a general aesthetic change from corrosion that creates maintenance to be done and does damage to the structure. Local or global.
- Corrosion maintenance does not eliminate all the corrosion damage and cracking or the repair is specified improperly or executed improperly thus creating a damage state not accounted for in the design. Local or global.
- Generation of a damage state that alters either the durability phase of life or the damage tolerant assessment of the structure or both.

- Creation of a widespread corrosion damage (WCD) state or a state of corrosion that impacts the occurrence of widespread fatigue damage (WFD) and its concomitant effects. [1, 3, 4, 13, 15, 25–27,31–36].
- Produce a condition that may cause a loss of fail safety in conjunction with one or more issues noted above.

The question of whether corrosion, corrosion fatigue, corrosion/ fatigue, and or stress corrosion cracking are safety concerns or just maintenance/economic concerns has been a point of discussion related to aircraft structural integrity for over 50 years. Nonetheless, a great deal of the aircraft structural integrity community believes that corrosion-related degradation is just an economic or maintenance concern. The issue of type of corrosion and its effects on structural integrity has been addressed in other summaries. This brief introduction gives a summary of some of the compilations of information related to corrosion in general. The major section that follows presents more information on the studies to date that have focused or are focusing on intergranular attack and exfoliation corrosion.

It was with the issue of safety or economic concerns that led Campbell and Lehay [12] and Wallace et al. [13] to pursue the presentation of technical facts and knowledge to illustrate the potential for a safety issue as well as maintenance and/or economic issue. Finally, Hoeppner et al. [27] reviewed failure data obtained from USAF, USN, USA, FAA, and the NTSB related to aircraft incidents and accidents in the USA from 1975–1994 to evaluate further the potential for corrosion and fretting-related degradation to be significant safety issues. A quote from the introduction to the paper [27] follows:

"On July 25, 1990, a pilot and crew were killed when the right wing outboard of the engine nacelle separated from their Aero Commander (now Twin Commander) 680 while performing a geological survey. The aircraft entered an uncontrolled decent and crashed into a field near Hassela, Sweden. Investigations revealed that the wing failed due to corrosion pits, which nucleated fatigue cracks in the lower spar cap" [27]. Although the accident occurred in Sweden, this accident sparked inspections of other Twin Commander aircraft worldwide. In November of 1991, Twin Commander released a service report detailing extensive cracking problems found in the lower spar cap of a US registry airplane. The Australian Civil Airworthiness Authority (CAA), on behalf of the Federal Aviation Administration (FAA), conducted fractographic analyses on ten cracks found in the component. The CAA determined that the cracks formed by intergranular attack, pits and resulted in stress corrosion cracks and that further extension occurred by fatigue. These failures will be referred to later in the section on IG and exfoliation and more detail will be provided. The mechanism overlap (two or more corrosion or degradation mechanisms being involved in change of damage state) frequently has been observed as documented by [13] as well.

The above example illustrates how corrosion pits and IG attack can severely jeopardize the structural integrity and safety of aircraft. In addition to corrosion, fretting and fretting fatigue have proven, on occasion, to be significant safety hazards. This paper will not deal with fretting and fretting fatigue as the first author has written extensively about this elsewhere and these mechanisms of degradation were not to be included in this brief summary.

Although the aircraft industry directs a great deal of attention to safety concerns, for many years it has relegated corrosion and fretting to maintenance, economic, and inspection issues. While the industry has developed some corrosion/fretting prevention programs, it has not done what it possibly should to quantitatively evaluate the effects of corrosion/fretting on structural integrity. What attempts have been made in this area appear to be sporadic and limited in number [27].

Walter Schütz addressed this issue further in the Plantema lecture at ICAF [25]. Furthermore, anyone that doubts the potential catastrophic consequences of corrosion-related degradation of aircraft structure would be assisted by reading Steve Swift's insightful presentation related to "The Aero Commander Chronicle" [26]. As a part of a technical paper delivered by Hoeppner et al. at ICAF-1999, they found the following with regard to pitting corrosion and pitting corrosion fatigue as listed in Table 2.

Table 2: Incidents from pitting corrosion and corrosion fatigue

Aircraft	Location of failure	Cause	Incident severity	Place	Year	From

Bell Helicopter	Fuselage, longeron	Fatigue, corrosion and pitting present	Serious	AR	1997	NTSB
DC-6	Engine, master connecting rod	Corrosion pitting	Fatal	AK	1996	NTSB
Piper PA-23	Engine, cylinder	Corrosion pitting	Fatal	AL	1996	NTSB
Boeing 75	Rudder control	Corrosion pitting	Substantial damage to plane	WI	1996	NTSB
Embraer 120	Propeller blade	Corrosion pitting	Fatal and serious, loss of plane	GA	1995	NTSB
Gulfstream GA-681	Hydraulic line	Corrosion pitting	Loss of plane, no injuries	AZ	1994	NTSB
L-1011	Engine, compressor assembly disk	Corrosion pitting	Loss of plane, no injuries	AK	1994	NTSB
Embraer 120	Propeller blade	Corrosion pitting	Damage to plane, no injuries	Canada	1994	NTSB
Embraer 120	Propeller blade	Corrosion pitting	Damage to plane, no injuries	Brazil	1994	NTSB
Mooney Mooney 20	Engine, interior	Corrosion pitting, improper approach	Minor injuries	TX	1993	NTSB
C-130	Bulkhead "Pork chop" fitting	Fatigue, corrosion pitting	Pressurization leaks	—	1995	LMAS
C-141	FS998 main frame	Corrosion pitting, stress corrosion cracking	Found crack during inspection	—	1991	LMAS

The examples shown in the table, taken with the general information cited in the references, clearly show that corrosion-related degradation is a significant safety issue in the assurance of structural integrity of aircraft. No such compilation has been done for exfoliation alone but needs to be done in the authors' opinion.

In recent years more emphasis has been placed on this issue of corrosion effects on structural integrity-especially after the fleet surveys subsequent to the Aloha Airlines accident (AA243) in 1988 [16]. Even though the NATO-AGARD community authorized the production of a manual on corrosion case studies and a great deal of information was presented in the manual published by AGARD [13], it is essential that the RMS deficiencies that may arise before accidents occur be recognized. This clearly has not been the case in all major fleets of aircraft whether they are military or commercial [12, 16, 19–22, 25–27]. Another issue that is clear is that deficiencies in the analysis of failures and the databases exist [27, 28].

The potential regrettable occurrence of accidents from corrosion-related crack formation/nucleation is a constant threat to aircraft safety. The following quote from the recent NATO RTO conference on fatigue in the presence of corrosion adds some understanding to the need for greater effort to understand the potential role of effects of corrosion on structural integrity.

Some of the workshop papers discussed the significance of corrosion-fatigue as a safety issue or an economic issue. There is ample data to support the contention that it is definitely an economic issue. There is also ample data to support the contention that it has not been a significant safety problem. However, the problem is certainly a potential safety concern if maintenance does not perform their task diligently. In addition, management must continuously update established maintenance and inspection practices to address additional real-time degradation threats for aircraft operated well beyond their initial design certification life. The economic issue alone is sufficient to motivate the support of research and development that can reduce the maintenance burden. This research will also reduce the threat of catastrophic failure from the corrosion damage.

(Lincoln, J., Simpson, D., Introduction to [36]).

Another quote from a different reference sheds further light on this issue [33, page 1-1].

At the present time, structural life assessments, inspection requirements, and inspection intervals, are determined by Durability and Damage Tolerance Assessments (DADTAs) using fracture mechanics crack growth techniques in accordance with the Aircraft Structural Integrity Program (ASIP). These techniques do not normally consider the effects of corrosion damage on crack initiation or crack growth rate behavior. Also, these techniques do not account for multiple fatigue cracks in the DADTAs of the structural components susceptible to WFD.

For aircraft that are not expected to have significant fatigue damage for many years, such as the C/KC-135, this approach has severe limitations since it does not account for corrosion damage or WFD. The impact of corrosion damage and WFD on stress, fatigue life, and residual strength must be understood to ensure maintenance inspections and repair actions are developed and initiated before serious degradation of aircrew/aircraft safety occurs.

Thus, the community clearly now recognizes the potential impact of corrosion-related degradation on structural integrity of aircraft. The need to understand the potential for the occurrence of corrosion on aircraft components is critical. Thus, to even begin the assessment of this potential the community needs to know the following:

- the chemical environment likely to be encountered on the structure of interest at the location of interest,
- the material from which the component is manufactured,
- the orientation of the critical forces (loads) applied externally and internally with respect to the critical directions in the material,
- the susceptibility of the material to occurrence of a given type of corrosion,
- the temperature of exposure of the component,
- the type of forces applied (i.e., sustained force or cyclic force-constant amplitude loading or variable amplitude loading),
- the type of exposure to the chemical environment (i.e., constant, intermittent), concomitant with the forces (corrosion fatigue or stress corrosion cracking) or sequentially with force (corrosion/fatigue or corrosion-fatigue),
- the rates of corrosion attack,
- the potential influence of the effects of corrosion on fatigue crack nucleation and propagation,
- the impact of any related corrosion degradation to residual strength,
- the potential for widespread corrosion damage to occur (WCD),
- the potential impact of corrosion on the occurrence of widespread fatigue damage (WFD) and its impact on structural integrity.

Obviously this is a formidable list but the assessment of these items is possible to some degree to make the estimation of the effects of corrosion more accurate than they have been to date.

This paper deals with the identification of the issues to be dealt with in establishing methods of estimating (predicting) the effects of corrosion To do this, various models are employed to be able to identify methods of establishing those components most susceptible to the ravages of corrosion.

CORROSION IN AIRCRAFT STRUCTURAL ALUMINUM ALLOYS

General

Corrosion is an electrochemical reaction process between a metal or metal alloy and its environment [37]. For corrosion to occur, four conditions must exist, namely, an anode, a cathode, an electrolyte, and an electrical path (flow of electrons). The anode and the cathode could be of two dissimilar metals or anodic and cathodic cells could be formed in the same metal alloy because of the potential difference in the constituent chemical elements or grain interior and grain or phase boundaries. Moreover, depending on the availability of oxygen (differential aeration cells) and electrolyte (differential concentration cells) on the surface of the metal alloy, special types of localized corrosion could occur. 2xxx (Al-Cu alloys) and 7xxx (Al-Zn alloys) series aluminum alloys are most commonly used in manufacturing aircraft structural components. This is currently true and has been true for some time. Depending upon strength and toughness requirements, different types of aluminum alloys such as 2024, 7075, 7178, and many others are used for commercial and military aircraft fuselage skins, wing skins, and other extrusions and forging such as stringers and fuselage frames. In general, 2024-T3 is used for skins and 7075-T6 for stringers and frames although many applications of these and other alloys in the 2xxx and 7xxx families exist. Lap or butt splices are the common configuration for longitudinal joints whereas butt joints are for circumferential joints. A common joining method is riveting and in some cases it is in combination with adhesive bonding. In older aircraft, spot welding also can be found. As Wallace and Hoeppner mentioned in their AGARD report on "Aircraft Corrosion: Causes and

Case Histories," in the initial stages, corrosion is in the form of filiform or pitting in the interior and exterior of fuselage skins [38]. Moreover, as noted in their report, crevice corrosion between the riveted sheets in fuselage joints is a significant issue and it is usually associated with the trapped small "stagnant solution." Furthermore, depending upon the chemical conditions this could lead to a combination of pitting, galvanic, or exfoliation corrosion. As well, it is recognized that fretting corrosion/wear in faying surfaces and within fastener holes plays a role in the corrosion mechanisms within aircraft joints [38]. The process of corrosion may start early in the process of manufacturing and continues when the aircraft enters its service. Therefore, it has been realized that the corrosion prevention and control program (CPCP) should be planned concurrently from the initial design until the aircraft is out of service. Furthermore design allowables should be established as with other major integrity issues.

Many types of corrosion mechanisms such as intergranular, exfoliation, pitting, crevice, fretting, microbiologically influenced corrosion, stress corrosion cracking, and hydrogen embrittlement have been found to occur in aircraft structural aluminum alloys [38]. Moreover, the synergistic effects of corrosion and the loading conditions have been found to initiate the corrosion fatigue failure process and the stress corrosion cracking failure process of aluminum alloy aircraft structural components. As identified, recently, in a report by the National Research Council's National Materials Advisory Board [39], corrosion in aircraft structural joints would result in the following: (i) significant changes in the applied stress because of material loss as well as corrosion product buildup that may cause "pillowing" or bulging of aluminum alloy sheet, (ii) hydrogen embrittlement that may result in reduced toughness, strength, and ductility of the material, and (iii) increase in fatigue crack growth rates that may severely hamper the planned inspection intervals. These issues have been discussed in workshops presented for the US-FAA and UCLA as well as FASIDE Int. Inc. workshops since 1971. In addition, the first author has frequently discussed the following other potential effects of corrosion on structural integrity:

- production of localized stress concentrations that act as crack nucleation sites,
- change of the structurally significant item (SSI),

- Modification of the fail safety by any of the above.

Moreover, recently, an attempt has been made to model loss of thickness due to crevice corrosion growth in a corroded lap joint.

Several metallurgical, mechanical, and environmental factors influence the corrosion process in aluminum alloys [40]. Metallurgically induced factors include heat treatment, chemical composition of alloying element, material discontinuities such as the presence of voids, inclusions, precipitates, second-phase particles, and grain boundaries as well as grain orientation. Environmental factors include temperature, moisture content, pH, type of electrolyte, and the time of exposure. Aircraft often are exposed to both external and internal environments. External surfaces of the aircraft are exposed to a variety of environments including rain, humidity, acid rain, deicing fluid, industrial pollutants, hot and cold temperatures, dust, high content of deposits of exhaust gases from engines, and salts. In addition, the inside of the aircraft is affected by condensed moisture, spilled beverages, cargo leak, deicing fluid, lavatory seepage, and accumulated water in the fuel tank, and others. Moreover, aircraft are exposed to wide ranges of environment depending upon their route and geographical location, namely, tropical, marine, industrial, and rural [38, 51].

In both military and commercial aircraft, internal and external wing structures as well as the fuselage bilge areas and flight control surfaces are found to be most affected by corrosion in a marine and tropical environment [41]. The major causes of corrosion in aging aircraft as observed in Indonesian aging aircraft were found to be due to spillage of toilet liquid, contamination due to spillage or evaporation from the cargo compartment, and contamination due to high humidity [42]. In addition, in these aircraft, corrosion was often found in the area surrounding the cargo compartment, wing structure, and landing gear. The types of corrosion found in these aircraft were of exfoliation, galvanic, filiform, and stress corrosion, and among these exfoliation corrosion was found in most cases [42].

Several "structural issues" such as exfoliation, pitting, stress corrosion cracking, fatigue cracking, fastener corrosion, wear, fatigue and corrosion, delamination and disbonds have been observed in the US Air Force aging aircraft as shown in Table 3 [39]. For example, in C/KC 135 fleet, crevice corrosion in the spot welded lap joint/doubler and corrosion around the steel fasteners on the upper wing skin have been recognized as significant corrosion issues [43]. In the later case, as was noted, there was a possibility of moisture from condensation or deicing solution trapped around fastener heads forming a galvanic couple. This was observed to result in intergranular attack of the grain boundaries leading to exfoliation [43].

Table 3: Corrosion and fatigue issues in the US Air Force aging aircraft

	Type of aircraft	Issues
1	C/KC 135 (Tanker aircraft)	Corrosion between fuselage lap joints and spot welded double layers, around fasteners in the 7178-T6 aluminum upper wing skins, between wing skins and spars, between bottom wing skin and main landing gear trunnions, between fuselage skin and steel doublers around pilot windows,
		Stress Corrosion Cracking (SCC) of large 7075-T6 aluminum forging (fuselage station 620, 820, and 960), corrosion and SCCof fuselage station 880 and 890 floor beams, wing station 733 closure rib, and corrosion in the E model engine struts.
2	C-141B (Transport aircraft)	Widespread Fatigue Damage (WFD) in the fuel drain holes in the lower surfaces of the wings, corrosion and SCC in the upper surface of the center wing, fatigue cracking and SCC around the wind shield, fatigue cracking in the stiffeners in the aft pressure door, SCC in the fuselage main frames, and corrosion in the empennage.
3	C-5 (Airlifter)	SCC of the 7075-T6 aluminum mainframes, keel beam, and fittings in the fuselage, 7079-T6 fuselage lower lobe and aft upper crown.
4	B-52H (Bomber)	Cracking in the bulkhead at body station 694, fatigue cracking in flap tracks and in the thrust brace lug of the forward engine support bulkhead, cracking in the side skin of the pressure cabin, aft body skins, and upper surface of the wing.

5	F-15 (Fighter aircraft)	Low-cycle fatigue cracking in the upper wing surface runouts, upper wing spar cap seal grooves, front wing spar conduit hole, upper in-board longeron splice plate holes, corrosion in nonhoneycomb structure including fuselage fuel tank, the outboard leading-edge structure of the wings, and the flap hinge beam.
6	F-16 (Fighter aircraft)	Cracking of the vertical tail attachment bulkhead at fuselage station 479, fuel vent holes of the lower wing skin, the wing attach bulkhead at fuselage station 341, the upper wing skin, fastener problems on the horizontal tail support box beam, and the ventral fin.
7	A-10 (Attack aircraft)	Fatigue cracking in the wing auxiliary spar cutout of the center section rib at wing station 90, outer panel front spar web at wing station 118 to 126, outer panel upper skin at leading edge. Fatigue cracking in the center fuselage forward fuel cell floor at the boost pump, forward fuselage gun bay compartment, forward fuselage lower longeron and skin at fuselage station 254, and center fuselage overwing lower floor panel stiffeners. Fatigue cracking in the aft nacelle hanger frame, thrust fitting and the engine inlet ring assembly skin/frame. Fatigue cracking in the main landing gear shock strut outer cylinder. Exfoliation corrosion in the 2024-T351 aluminum lower wing skin, 7075-T6 aluminum upper wing at the leading edge, 2024-T3511 aluminum lower front spar cap, 7075-T6 aluminum fuselage bottom skin 2024-T3/7075-T6 aluminum fuselage side skin and beaded pan, and 2024-T3511 aluminum horizontal stabilizer upper spar caps. Pitting corrosion in the 9Ni-4Co-0.3C steel wing attach fitting bushing and lug bore, main landing gear fitting attach bolts, 7075-T6 aluminum aft fuel cell aft bulkhead, and 2024-T351 center fuselage upper longeron. SCC in the wing attach bushing flange, and the main landing gear attach bolts.

8	E-3A (Airborne Warning and Control System)	Fatigue and corrosion in the 7178-T6 rudder skins and spoiler actuator clevis. Exfoliation corrosion in the 7178-T6 upper wing skin, leading edge slats, main landing gear door, fillet flap, fuselage stringer 23, and magnesium parts. Delaminationand disbonds in the windows, floor panels, and nose radome core. Wear in the antenna pedestal turntable bearings.
9	E-8 (Joint Surveillance and Attack Radar System)	"Small" fatigue cracks in fastener holes in the 7075-T6 aluminum stringers, in the 2024-T3 aluminum skins.
10	T-38 (Air training command aircraft)	Fatigue cracking in the lower surface of the wing, lower wing skin fastener holes, wing skin access panel holes, milled pockets on the lower wing skin, and the fuselage upper cockpit longerons. SCC in the fuselage cockpit upper and lower longerons, fuselage forgings. Honeycomb corrosion in the horizontal stabilizer (due to water intrusion), and the landing gear strut door.

Examination of C/KC 135 fuselage lap splices (stiffened aluminum lap joint) revealed that outer skin corrosion was predominantly intergranular and exfoliation [44]. Moreover, extensive cracking was noted at these sites in the outer skin. In addition, extensive "pillowing" with more than 300% change in volume due to corrosion products along the faying surfaces was observed. In the rivet/shank region, severe localized corrosion and intergranular corrosion were observed. The fracture of rivet heads was attributed to high local stress due to environmentally assisted cracking at the junction. As well, in this study, solution samples were collected from selected areas of lap splice joints and the solution analysis showed the presence of several cations such as Al^{3+}, Ca^{2+}, Na^+, K^+, and Ni^{2+} and also anions Cl^-, SO_4^{2-}, NO_3^- and . Subsequent potentiodynamic tests using solution containing these ions led to the belief that dissolution rates could completely penetrate the fuselage outer skin during service life [44].

Thus, in addition to fatigue cracking, different corrosion mechanisms occur in aircraft structures depending upon their location, geometry, exposure to environment, and loading conditions. Research studies conducted within the Quality and Integrity Design Engineering Center (QIDEC) at the University of Utah as well as other related studies are briefly discussed below.

Intergranular and Exfoliation Corrosion

Exfoliation corrosion is believed to be a manifestation of intergranular corrosion. Intergranular corrosion results from either the segregation of reactive impurities or from the depletion of passivating elements at the grain boundaries. This makes the regions at or surrounding the grain boundaries less resistant to corrosion resulting in preferential corrosion. The high strength aluminum alloys such as 2xxx and 7xxx series are highly susceptible to intergranular corrosion [37]. Exfoliation corrosion is a form of intergranular attack that occurs at the boundaries of grains elongated in the rolling direction. The 7xxx series aluminum alloys are particularly less resistant to exfoliation corrosion because during heat treatment (to achieve maximum desirable strength) their constituent elements copper and zinc accumulate at grain boundaries leaving the adjacent region free of precipitates. As aluminum and aluminum intermetallic compounds are highly reactive in the EMF series and aluminum is anodic to copper in the galvanic series, the resulting galvanic couples cause the grain boundaries to preferentially corrode (intergranular attack). McIntyre and Dow have related the localized corrosion problems in the 7075-T7352 fuel tanks of underwater weapon systems to intergranular corrosion [45]. In their study, aluminum alloys 7075 and 6061 were exposed to artificial seawater containing nitrate ions. It was observed that accelerated intergranular corrosion occurred in 7075 alloys. From the test results, they hypothesized that refueling the improperly cleaned fuel tank may cause the propellant in contact with the small quantity of sea water remaining in the fuel tank resulting in the release of nitrate ions from a hydrolysis process leading to reduced pH that may cause the dissolution of the oxide film (localized corrosion). They further hypothesized that corrosion eventually propagates to the bulk regions of the alloy due to intergranular attack by the preferential corrosion of reactive $MgZn_2$ intermetallic compounds located at grain boundaries. This was found to be true for 7075 aluminum alloy but not for 6061 aluminum alloy because the latter does not contain either Cu or Zn as alloying element [45].

Reducing the impurities such as iron and silicon as well as heat treatment modifications in aluminum alloys have resulted in an increase in the resistance to exfoliation corrosion [46]. For example, overaged 7075-T7 alloys are more resistant to exfoliation corrosion when compared to 7075-T6 alloys. In addition, Rinnovatore showed that in the T6 temper, exfoliation corrosion resistance was found to be greater for forgings produced from rolled bar stock than forgings from extruded bar stock [47]. Moreover, it was shown that rapid quenching from the solution temperature in cold water increased exfoliation corrosion resistance of forgings tempered to T6.

Fatigue and exfoliation interactions have been studied. Mills reports that most of the studies have been performed during the last five years on this issue although Shaffer in 1968 reported significant reduction in the fatigue life of exfoliated extruded 7075-T6 spar caps [48]. Moreover, multiple crack nucleation sites were observed in 7075-T651 [49] and 2024-T3 [50] aluminum alloy specimens when the specimens were subjected to exfoliation corrosion and then fatigue tested under positive R values with constant amplitude loading. Mills found an 88% decrease in the fatigue life of the specimens with prior exfoliation corrosion damage when compared to specimens tested without prior-corrosion damage. Chubb et al. showed in their study using panels containing fastener holes that the end grains exposed in the rivet holes would be the potential corrosion sites that could eventually result in multiple site damage.

In a recent study [48], experiments were performed to determine the effect of exfoliation on the fatigue crack growth behavior of 7075-T651 aluminum alloy. First the specimens were subjected to prior-corrosion damage using ASTM standard EXCO corrosive solution and then fatigue tested in corrosion fatigue environments of dry air, humid air, and artificial acid rain. Test results indicated that prior-corrosion damage resulted in higher crack growth rates than when tested in dry air as well as in acid-rain environments when compared to uncorroded specimens. Fractographic analysis showed quasi-cleavage fracture close to the exfoliated edge of the specimens tested in all the three environments indicating embrittlement by prior corrosion. Thus, embrittlement by prior corrosion was stated to "result in accelerated crack nucleation, faster short crack growth, and earlier onset of fatigue phenomena such as multiple site damage."

Corrosion Fatigue

Corrosion fatigue is defined as "the process in which a metal fractures prematurely under conditions of simultaneous corrosion and repeated cyclic loading at lower stress levels or fewer cycles than would be required in the absence of the corrosive environment" [40]. Corrosion acting conjointly with fatigue can have major effects on materials in structures of aircraft. First, corrosion can create discontinuities (pits, cracks, etc.) that act as origins of fatigue cracks with significant reductions in life at all stress levels. In crack propagation, corrosion effects are well known to produce accelerated fatigue crack propagation. The combination of aggressive environment and cyclic loading conditions has been observed to accelerate crack growth rates in aluminum alloys. Several mechanisms were proposed to explain the corrosion fatigue process [37]. They are (i) dissolution of material at the crack tip in corrosive environment, (ii) hydrogen embrittlement in which diffusion of hydrogen (a byproduct of corrosion process) into the lattice space could weaken the atomic bonds thereby reducing the fracture energy, (iii) theory of adsorbed ions in which the transport of critical species to the crack tip results in lowering of the energy required for fracture, and (iv) film-induced cleavage in which it is hypothesized that crack speed would increase at the film-substrate interface when the crack grows through the low-toughness oxide layer leading to the rupture of the film.

In general, corrosion fatigue effects on crack propagation are more pronounced at lower stress intensities whereas at higher stress intensities the crack propagates at such a high rate that the effects of chemical dissolution or localized embrittlement will be negligible. Several parameters affect corrosion fatigue crack propagation rates. For example, crack growth rates increase with increase in the stress intensity range. Also, at lower frequency corrosion fatigue effects will be more severe than at higher frequency because of the time-dependent nature of the process. Increase in R value has been found to generally increase corrosion fatigue crack propagation rates. As well, increasing the concentration of corrosive species, lowering the pH, increasing the moisture content, and temperature usually result in more severe effects [40].

The most common corrosion fatigue environment that is simulated in laboratory testing is 3.5% NaCl as it is believed to result in severe general corrosion rates and it represents roughly the salinity of sea water. In addition, other environments such as humid air, salt sprays, and artificial acid rain (to simulate industrial pollutants) also are used to characterize corrosion fatigue crack growth behavior of aluminum alloys. As aircraft are exposed to several complex chemical environments both inside and outside, no single environment could simulate the actual condition. Therefore, a few studies used sump tank water that was considered close to a "realistic chemical environment" [7]. The quest for realistic corrosion fatigue environment led Swartz et al. [51] to collect and analyze solution samples from bilge areas, external galley, and lavatories of five different airplanes. As a result, a new chemical environment was developed to perform corrosion fatigue crack growth experiments on 2024-T351, 2324-T39, 7075-T651, and 7150-T651 aluminum alloys. For all the alloys studied the fatigue crack propagation rates in synthetic bilge solution were found to be between the dry air and the 3.5% NaCl data. In another study [52], cyclic wet and dry environment was simulated in characterizing the corrosion fatigue crack growth rates in 2024-T351 aluminum alloy. It was hypothesized that during the dry cycle the partial evaporation of the aqueous solution may allow some chemical species to get deposited at the crack tip, and then in the wet cycle when the rehydration occurs, corrosion could occur at a greater rate than before.

To simulate aircraft service corrosion, fatigue crack growth studies were conducted on service corroded 2024-T3 aluminum panels extracted from a C/KC-135 aircraft [53]. Test results showed that in some cases fatigue crack growth rates were two or three times greater in the corroded material, however, in other cases, there was little difference. It was observed that "the difference in the crack growth rates was due to high variability in the amount of corrosion damage between specimens."

Corrosion Pillowing and Its Effect on Structural Integrity of Aircraft Lap Joints

Recently, some studies have shown that the increase in stress levels is not only because of the thickness loss due to corrosion but also due to the volume of the corrosion product buildup in a joint [54]. Also, evidence shows that lap joints contain "faying cracks" under the rivet heads in the corroded areas. The complexity of this issue as explained by Komorowski et al. is that "the majority of the cracks had not penetrated the outer skin surface and appeared to grow more rapidly along the faying surface creating a high aspect ratio semi-elliptical crack and it is difficult to detect and affects the structural integrity of the joint" [54]. As reported by Komorowski et al. [54], the major corrosion product in the lap splices is found to be aluminum oxide trihydrate, an "oxide mix" which has a high molecular volume ratio to the alloy. As the oxide is insoluble, it is found to remain within the joint and in turn is responsible to deform the skins in the joint which usually gives a bulging appearance, commonly termed as "pillowing." Moreover, finite element analysis revealed that for a two-layer joint, the stresses due to 6% thinning due to corrosion resulted in stress more than the yield strength of 2024-T3 aluminum alloy [55]. In addition, "pillowing-induced deformation" was observed on the corroded joints after removal of the rivets and the separation of the skin. Moreover, multiple cracks were found to nucleate from rivet holes. Fracture mechanics analysis has shown that as the pillowing increases, the stress intensity factor for the crack edge along the faying surface increases [55]. On the other hand, the stress intensity factor decreases for the crack edge along the outer surface. Therefore, it was hypothesized that pillowing produces compressive stresses in the rivet area on the outer surface because of the resultant bending stresses. At the same time, high tensile stress is produced on the faying surface resulting in more rapid growth of faying surface cracks in the direction of the row of rivets than through the skin towards the outer surface [55].

Pitting Nucleation Theories

Pitting corrosion is defined as "localized corrosion of a metal surface, confined to a point or small area that takes the form of cavities" [40]. Pitting is a deleterious form of localized corrosion, and it occurs mainly on metal surfaces which owe their corrosion resistance to passivity. The major consequence of pitting is the breakdown of passivity, that is, pitting, in general, occurs when there is breakdown of surface films when exposed to pitting environment. Pitting corrosion is so complicated in nature because "oxide films formed on different metals vary one from another in electronic conduction, porosity, thickness, and state of hydration" [56]. The empirical models that have been developed to understand the pitting process are closely related to the integrity of the metal oxide film. The salient features of the empirical theories related to pit nucleation mechanisms are mentioned in Table 4.

Table 4: Pit nucleation theories

Proposed by	Theory
Evans [57] (1929-30)	Proposed penetration theory. Ability of a chloride ion to penetrate the film was linked to the occurrence of pitting. Halide ions are assumed to be transported from the film-solution interface to the metal-oxide interface either by the application of electric field or exchange of anions.
Hoar [58] (1967), Hoar and Wood [59] (1960s)	Assumed the adsorption of anions on the oxide surface as the key aspect in the pit nucleation process. Proposed "ion-migration" model that involves activating anions that enter the oxide film lattice without exchange thereby increasing the ionic conductivity of the film resulting in local high anodic dissolution rates and pitting. Proposed "mechanical" model in which it was assumed that adsorption of anions at the oxide-solution interface lowers the interfacial energy resulting in the formation of cracks in the protective oxide film under the influence of the "electrostatic repulsion" of the adsorbed anions. Suggested a concept of local acidification of pit as a critical factor in pit growth.
Bohni and Uhlig [60] and Kolotyrkin and Ya [61] (1961–1967)	Proposed adsorption theory in which at a certain value of the potential (pitting potential) the adsorption of aggressive anions on the metal surface displaces the passivating species such as oxygen. Kolotyrkin suggested that adsorption of anions at preferred sites forming soluble complexes with metal ions from the oxide. Once such species leave the oxide, thinning of the film starts locally increasing the electric field strength which accelerates the dissolution of the oxide.

Sato [62–64] (1971, 1982)	Proposed that at a critical potential an internal film pressure exceeds the critical compressive stress for film fracture. Considered thinning of film at local sites and suggested that pitting occurs only when a critical concentration of aggressive anions and a critical acidity is locally built up.
Lin et al. [65] (1981)	Proposed that metal vacancies may accumulate as a result of the diffusion of metal cations from the metal/film to the film/solution interface, forming voids at the metal/film interface. When the voids grow to a critical size the passive film will collapse leading to pit growth.

Therefore, nucleation of pits generally involves certain localized changes in the structure and properties of the oxide film. However, propagation of pits is related to the dissolution of the underlying bulk metal. Further discussion on this subject is presented later in this paper.

PITTING CORROSION

Overview

Pitting is classified as a localized attack that results in rapid penetration and removal of metal at "small" discrete areas [68]. An electrolyte should be present for pitting to occur. The electrolyte could be a film of condensed moisture or a bulk liquid. How and when pitting occurs on a metal depends on numerous factors, such as type of alloy, its composition, integrity of its oxide film, presence of any material or manufacturing-induced discontinuities, and chemical and loading environment, to name a few. Many metals and their alloys are subject to pitting in different environments. These include alloys of carbon steels, stainless steels, titanium, nickel, copper, and aluminum [69].

In passivated metals or alloys that are exposed to solutions containing aggressive anions, primarily chloride, pitting corrosion results in local dissolution leading to the formation of cavities or "holes." The shape of the pits or cavities can vary from shallow to cylindrical holes, and the cavity is approximately hemispherical [70]. The pit morphology depends on the metallurgy of the alloy and chemistry of the environment and the loading conditions. As observed first by McAdam and Gell in 1928 [71], these pits may cause local increase in stress concentration and cracks may nucleate from them [71].

According to Foley [72], pitting corrosion of aluminum occurs in four steps: (1) adsorption of anions on the aluminum oxide film, (2) chemical reaction of the adsorbed anion with the aluminum ion in the aluminum oxide lattice, (3) penetration of the oxide film by the aggressive anion resulting in the thinning of the oxide film by dissolution, and (4) direct attack of the exposed metal by the anion.

The susceptibility of a metal to pitting corrosion as well as the rate at which pitting occurs on its surface depends on the integrity of its oxide film. Therefore, a brief overview of the mechanisms of the formation of passive film is discussed below.

Formation of Passive Films and Their Growth

The following discussion on the oxide film formation and its growth is extracted from [73].

Early investigators examined the effects of natural waters on metals by placing them outside. One investigator, Liversidge, in 1895, observed that an aluminum specimen,

... "lost its brilliancy, and became somewhat rough and speckled with grey spots mixed with larger light grey patches; it also became rough to the feel, the grey parts could be seen to distinctly project above the surface, and under the microscope they presented a blistered appearance. This encrustation is held tenaciously, and does not wash off, neither is it removed on rubbing with a cloth" [74].

Liversidge proposed that a hydrated aluminum oxide had formed, but did not confirm this with further testing of the layer. He did, however, note that when weighed, the aluminum specimens gained weight with exposure, rather than losing weight [75]. It was later confirmed that the weight gain was due to formation of an oxide film [76]. Although Liversidge suggested the formation of an aluminum oxide film, subsequent investigators proposed other theories to explain the passive behavior of aluminum. Some of these were changes in the state of electric charge on the surface, changes in valence at the surface, and a condensed oxygen layer [77]. Dunstan and Hill proved the presence of the oxide film on the surface of the metals in 1911. Through experiments with iron, they determined that the passive film was reduced at 250°F, the temperature at which magnetic iron oxide is reduced. Similar films were found on other metals [77]. Barnes and Shearer attempted to determine the constitution of passive films on aluminum and magnesium in 1908. They determined that aluminum formed hydrogen peroxide when reacting with water and that the passive film consisted of $Al_2(OH)_6$ [76]. This was later determined to be incorrect [78].

Structure of the Passive Film in Aluminum

It later was determined that this film on aluminum consists of an aluminum oxide created when the aluminum comes in contact with an environment. Generally, this film is amorphous; however, under certain circumstances it will develop one of seven crystalline structures:

- gibbsite (also called hydrargillite): (α-$Al_2O_3 \cdot 3H_2O$),
- bayerite: (β-$Al_2O_3 \cdot 3H_2O$),
- boehmite: (α-$Al_2O_3 \cdot H_2O$ or $AlO \cdot OH$),
- diaspore: (β-$Al_2O_3 \cdot H2O$),
- gamma alumina: (γ-Al_2O_3),
- corundum: (α-Al_2O_3)
- combinations of aluminum oxides with inhibitors, for example, ($2Al_2O_3 \cdot P_2O_5 \cdot 3H_2O$).

Gibbsite and diaspore structures are not found during corrosion of aluminum, but are frequently found in bauxite ores. Boehmite, bayerite, gamma alumina, and corundum are sometimes found in the passive layers of aluminum under certain conditions. Additionally, bayerite is frequently found as a corrosion product during pitting of aluminum. Combinations of aluminum oxides with inhibitors are not understood very well in the literature, but it is known that they will combine with oxide layer to form improved corrosion resistance through changing the passive film structure. Several researchers have studied changes in the amorphous structure of the oxide film. In one investigation, the passive film formed on the pure aluminum sheet revealed changes in structure with an increase in temperature and oxygen content. Prior to heating, the structure was reported to be amorphous oxide. As the temperature was increased, the amorphous film thickened, formed boehmite, and bayerite. The rate of film formation increased with temperature, and with an increase in oxygen content, intergranular attack began. The researcher suggested the following sequence of events in the formation: boehmite is nucleated at dislocation centers that are at the surface of the amorphous film; it then grows by a diffusion mechanism. During thickening of the boehmite, a process occurs that allows aluminum ions to escape into the solution, which results in bayerite growth [79].

Other investigations revealed that aluminum in the molten state would develop an oxide film of gamma alumina which will convert to corundum when exposed to dry air. Aluminum sheet in water at temperatures below 70 to 85°C after long aging will develop a passive film consisting of bayerite. Boehmite is found on aluminum exposed to water at high temperatures (above 70 to 85°C) [78]. More recently, researchers have found small regions of crystallized γ-alumina within the amorphous layers created during anodizing [80].

During exposure to air and water, alumina will form a passive film with a duplex structure. The film will consist of two layers, a permeable outer layer and a protective, nonporous layer next to the metal's surface. In the case of an air environment, the protective layer is thicker and the permeable layer is comparatively thin. In the case of an immersion in water, the permeable layer is thicker and the protective layer is thinner. In both cases, the total thickness of the duplex film is the same [78].

The protective layer will quickly reach maximum thickness, with the permeable layer growing slower. The growth rate of each layer depends on a few parameters. In air, it is dependent on temperature; in water, it is dependent on temperature, oxygen content, pH, and the type of ions present in the electrolyte; in anodization procedures, it depends on electrolyte and applied potential. The film is typically formed on pure aluminum when the pH of the solution is between 4.5 to 8.5 [78].

Other researchers have suggested that the permeable outer layer consists of hexagonal close-packed pores in pure aluminum. The size of these pores will depend on conditions of formation. Sealing processes in an attempt to improve the characteristics of the passive film sometimes control these conditions of formation. In sealing processes, the pores are blocked or made smaller by boehmite or gamma alumina formation, nickel acetate is added to obstruct the pores, and dichromates or chromates can be added to create pores of a different structure [59].

The passive film formed on metals will differ according to the environment in which it forms. Studies done by Seligman and Williams in the 1920s illustrate this difference. In experiments with tap water, the presence or absence of certain impurities caused either the passive film to breakdown and the metal to corrode or the film will become thick and less susceptible to corrosion. They determined that nitrates and chromates would combine with the passive film and serve to increase resistance of the passive film to localized corrosion [81]. Later studies emphasized this conclusion. One researcher found a film of 55,000 angstroms in distilled water and another found a film of only 4,800 angstroms for the same alloy (AA-1099) immersed in tap water [78]. Additionally, experiments performed by Bengough and Hudson on aluminum in sea water showed that the passive film varied with corroding liquid and with different alloying elements [75].

In a more recent paper, researchers determined that the reaction between aluminum and water takes place in three steps: formation of the amorphous oxide, dissolution of the oxide, and deposition of the dissolved products as hydrous oxide. In the first step, the amorphous oxide layer is formed and grows by the anodic and cathodic reactions present at the water/metal interface. The second step involves a hydrolysis reaction with the surface which depends on temperature, pH and aluminum concentration, and the last step is accomplished when the resulting hydroxide is deposited on the surface. The rate at which the film will grow is controlled by the diffusion of water molecules through the existing layers. At temperatures between 50 and 100°C, pseudoboehmite grows on the amorphous oxide. At 40°C, however, bayerite crystallization occurs and with time will overcome the pseudoboehmite [82].

Upon exposure of an air-formed film to water, the air-formed film will break down and another film will form that is thicker and contains more water. The rate at which the film is reformed depends on the anions present and the temperature [78]. In more recent work, the water in the aluminum passive film has been stated to be a medium for the mobilization for aluminum cations and deposited anions [83].

In air, the thickness of the passive film is dependent on humidity. In higher humidity, the oxide layer is thicker. The growth rate of the film, however, does not depend on humidity. Rosenfeld et al. found that in high purity air, the growth rate was not changed. However, when small amounts of impurities were added, growth was accelerated in humid air [84]. In addition to impurities, the growth of the film is highly dependent on temperature. Below 200°C, the film will grow only to a few hundred angstroms, above 300 to 400°C, the rate gradually increases, between 400 and 600°C, the film will grow to a thickness of 400 angstroms, at 450°C, the film will crystallize to gamma alumina [78].

Pitting Potential and Induction Time

According to Szklarska [69], the susceptibility of a metal or alloy to pitting can be estimated by determination of one of the following criteria:

- characteristic pitting potential,
- critical temperature of pitting,
- number of pits per unit area or weight loss,
- The lowest concentration of chloride ions that may cause pitting.

   One of the most important criteria to determine an alloy's susceptibility to pitting corrosion is to find the pitting potential, that is, the potential at which the passive film starts to break down locally. The potential above which pits nucleate is denoted by E_p and the potential below, which pitting does not occur and above which the nucleated pits can grow, is often indicated by E_{pp}. Once the passive film begins to breakdown, the time it takes to form pits on a passive metal exposed to a solution containing aggressive anions, for example, Cl^-, is called the induction time or incubation time [69]. The induction time is meaningful in a statistical sense as it represents the average rate of reaction over the whole surface to produce a measurable increase in current. It should not be considered as the time to form the first pit. This is because "micro" pits have been observed to form during the induction time [72]. The induction time is usually denoted by τ. It is measured as the time required producing an appreciable anodic current at a given anodic potential. It is expressed as $1/\tau = k' (E - E_p)$, where E is the applied potential and K is a function of Cl^- ion concentration [85]. In general, pitting potential decreases with increasing Cl^- ion concentration.

The most commonly used relation for estimating t is based on an exponential relationship between time and activation energy, that is, $1/\tau = Ae^{-Ea/RT}$; the activation energy needed for pit nucleation can be obtained from an Arrhenius plot of log $1/\tau$ versus 1/temperature [72]. As well, Hoar [58] has proposed a relationship $1/\tau = K(Me)^m(X-)^n$ to estimate the induction time. Where Me is the metal ion concentration, X^- is the halide ion concentration, and m and n are orders of reaction which are determined experimentally. Subsequent to the nucleation of pits it has been observed that they grow. The following subsection presents a discussion of pit growth.

Pit Growth Rate and Pit Morphology

Godard [86] developed a simple but effective relation based on the experimental data to estimate the rate at which pits grow. The empirical relation he developed was $d = K(t^{1/3})$.. Even though he found this relation when tested using aluminum, it was observed to be true for other materials in different types of water environments. In general, the rate of pit growth depends on several factors such as temperature, pH, properties of passive films, chloride ion concentration, presence of anions and cations in solution, and the orientation of the material [5]. The pit growth can be viewed as a direct interaction of the exposed metal with the environment.

Upon observing the geometry of the pits formed on 7075 aluminum alloy in halide solutions, Dallek and Foley [87] proposed a pit growth rate expression $i - i_p = a(t - t_i)^b$ in which current was expressed as a function of time. In this expression, i is the dissolution current, i_p is the passive current, t is the time, t_i is the induction time, a is the constant depending on the halide, and b is the constant depending on the geometry of the pit. From this expression, a plot of $\log(i - i_p)$ versus $\log(t - t_i)$ will give the slope b. Dallek predominantly observed pits of hemispherical shape. However, Nguyen and Foley [88] have observed hemispherical pits at low potential on 1199 aluminum alloy in chloride solutions and at high potential they observed a porous layer film covered on the pit mouth with orifice at the center. This study indicated the effect of potential on the morphology of pits. Chloride ion concentration also was found to affect the pit morphology. Baumgartner and Kaesche [89] observed that in dilute to medium concentrated solutions, pit morphology was "rough" whereas at high concentration, pits were found to be "smooth and rounded." In addition, a recent study by Grimes [73] showed clearly the effect of loading conditions on the morphology of pits. This study was conducted on 7075-T6 aluminum alloy in 3.5% salt water under three different loading conditions, namely, zero, sustained, and cyclic. It was found that the pits propagated under cyclic loads were three times larger in cross-sectional area when compared to those grown under sustained or zero load conditions. Also, it was found that most of the pits originated from the grain boundaries. This study concludes that the effect of both mechanical and chemical environment must be considered in pitting corrosion studies. However, when studying the effect of pitting on the fatigue life of aluminum alloy 7075-T6 in 3.5% NaCl solution, Ma [90] found that although the test frequency (5 and 20 Hz) had a pronounced effect on the total corrosion fatigue life, the fatigue test frequency did not have any effect on the pit morphology. On the other hand, Chen et al. [91] have found that the size of the pit from which a crack nucleated was comparatively larger at the lower frequencies and stresses than at higher frequencies and stresses when fatigue tested using 2024-T3 aluminum alloy.

Mechanisms of Pit Nucleation

In general, pit nucleation mechanisms are classified into three categories: (i) adsorption-induced mechanisms, (ii) ion-migration and penetration models, and (iii) mechanical film breakdown theories.

Adsorption-Induced Mechanisms

In this section mechanisms of pit nucleation based on the adsorption of aggressive anions at energetically favored sites are discussed. Many researchers including Uhlig et al. [60, 92, 93], Hoar [58], Hoar and Jacob [94], and Kolotyrkin [95] have suggested mechanisms related to the ion-adsorption concepts (see Table 5). Many of the mechanisms proposed in the literature consider this as a necessary step in the pit nucleation process. Uhlig [60, 92, 93] and Kolotyrkin [95] independently proposed that both oxygen and chlorine anions can be adsorbed onto the metal surfaces. When the metal is exposed in air, oxygen is adsorbed by the metal resulting in the formation of passive oxide film. Consequently, a chemical bond is established between the oxygen anion and the metal cation. This process is known in corrosion terminology as "chemisorption." Chemisorption results in the formation of a metal compound that covers the surface of a metal. If aluminum is exposed in oxygen, the resulting compound is aluminum oxide, that is, Al_2O_3. However, the type of compound that is formed on the metal surface depends on the environment in which the metal is exposed. For example, in the case of salt water, Cl^- ions in addition to oxygen are present. When oxygen is adsorbed, passivation of metal occurs whereas if chlorine anion is adsorbed, it does not result in passivation but breakdown of passivity occurs.

Table 5: Adsorption-induced mechanisms

	Proposed by	Summary	Description	Limitations
1	Uhlig et al. [92] 1950–69, Kolotyrkin 1961 [95], Hoar 1967 [58]	(i) Proposed concepts based on either competitive adsorption or surface complex ion formation.	(i) In competitive adsorption mechanism Cl– anions and passivating agents are simultaneously adsorbed. Above a critical potential Cl– adsorption is favored resulting in the breakdown of passivity. (ii) Kolotyrkin suggested that there were critical Cl–/inhibitor concentration ratios, depending on the potential above which pitting would occur.	Occurrence of induction times varying with passive film thickness cannot be explained.

| 2 | Sato 1982 [63,64] | (i) Proposed a theoretical concept based on the potential dependent transpassive dissolution which depends on the electronic properties of the passive film.

(ii) The electrochemical stability of a passive film depends strongly on the "electron energy band structure" in the film. | (i) Stated that the critical potential above which potential-dependent dissolution of the film occurs will be less noble at the sites of chloride ion adsorption.

(ii) As a result of the increased dissolution rate above the critical potential, local thinning of the passive films occurs until a steady state is reached.

(iii) Proposed that the local thinning of the oxide film as a mechanism of pit "initiation".

(iv) Included the effect of dislocations similar to the influences of Cl− ions. | Knowledge of the electronic properties of passive films has not been fully understood.

Experimental evidence for this mechanism is lacking. |

As proposed by Kolotyrkin [95], below the pitting potential, metals may prefer to adsorb oxygen, and above this critical potential metals may adsorb halides, such as Cl^-. This mechanism is termed "competitive adsorption" as the presence of different anions will compete with the oxygen to be chemisorbed by the metal. Therefore, at or above the pitting potential, chlorides and other aggressive anions if present combine with the metal and then diffuse from the metal's surface into the solution. Subsequently, it combines with water in solution to form metallic oxides, hydrogen and chloride ions. These chloride ions are attracted to the surface of the metal and the process begins again. It was hypothesized by many researchers that the chloride ions might diffuse to regions of high energy such as inclusion, dislocations, and other form of discontinuities.

Hoar [58] and Hoar and Jacob [94] originally proposed a "complex ion formation theory" which stated that the formation of Cl^- containing complexes on the film-solution interface might lead to a locally thinned passive layer. This was proposed because Cl^--containing complexes are more soluble when compared to complexes formed in the absence of halides. They assumed that a high-energy complex is formed when a small number of Cl^- ions jointly adsorb around a cation in the film surface, which can readily dissolve into solution. This creates a stronger anodic field at this site that will result in the rapid transfer of another cation to the surface where it will meet more Cl^- and enter into solution. Experimental support was provided for this concept by Strehblow et al. [96] by conducting an investigation on the attack of passive iron by hydrogen fluoride. They found that the breakdown process occurred with complete removal of the passivated oxide layer. It was observed that hydrogen fluoride catalyzed the transfer of Fe^{3+} and Ni^{2+} ions from oxide into the electrolyte. As mentioned in a paper by Bohni [97], similar observation was made in another study by Heusler et al. regarding the influence of chloride containing borate and phthalate solutions on the passive film breakdown of iron. Different behavior of Cl^- and F^- ions in the pit nucleation process was proposed in a model by Heusler et al. Cl^- ions were suggested to form only two dimensional "clusters" leading to the localized thinning of the passive layer. However, it was proposed that F^- ions adsorb homogeneously on the oxide surface thereby promoting a general attack. It should be noted that the proposed models did not take into account material discontinuities such as point "defects," dislocations, inclusions, voids, and others. Also, another model based on the concept of an increased probability of "electrocapillary film breakdown" was proposed by Sato [63, 64] (see Table 3). Although Sato includes the effect of dislocations in this purely theoretical approach, no experimental evidence was found in the literature to support his model. However, Sato's theoretical model proposed that n-type passive oxide films are more stable than p-type films because of the difference in the band structure of electron levels.

From these studies it can be concluded that in addition to chloride anions, other anions such as chromate and sulphate also get adsorbed changing the nature of the compound. In addition, as observed by Richardson and Wood [98], enhanced adsorption takes place at the "imperfections or flaws" in the oxide film. These discontinuities in the film usually become the sites of anion adsorption. Nilsen and Bardal [99] have observed by measuring the pitting potential of four aluminum alloys (99% pure Al, Al-2.7Mg, Al-4.5Mg-Mn, and Al-1Si-Mg) and found that the pitting potential values for the four alloys were within only 25 mv. From this study, they concluded that alloy composition does not directly depend on the adsorption step of the process.

Ion-Migration and Penetration Models

A few models (see Table 6) were proposed based on either penetration of anions from the oxide/electrolyte interface to the metal/oxide interface or migration of cations or their respective vacancies. This theory is based on the concept that Cl^- ions migrate through the passive film and results in breakdown of the film once they reach the metal/film interface. Hoar [58] explained that when a critical potential is reached, smaller ions, like Cl^-, may penetrate the film under the influence of an electrostatic field which exists across the film. These aggressive anions prefer the high energy regions like grain boundaries and impurities as sites for migration because these regions produce thinner passive films locally. During the migration, the ions either pass through the film completely or they may combine with the metal cation in the midst of the film resulting in the formation of what is called a "contaminated" film, which is a better conductor than the "uncontaminated" film. This process results in an autocatalytic reaction, which encourages more ions to penetrate the film. This hypothesis is supported by some researchers as they have observed a higher concentration of Cl^- ions over thin films on the surface of iron and that the time to breakdown the film increases with the thickness of the film [94]. It was further hypothesized that Cl^- ions first fill anion vacancies on the surface of the passive film and then migrate to the metal/oxide interface. However, other works revealed that the time required for Cl^- to penetrate through the film is much longer than the induction time measured experimentally [97].

Table 6: Ion-migration and penetration models

	Proposed by	Summary	Description	Limitations
1	Hoar et al. 1967 [58, 59]	(i) Presented that when the electrostatic field across the film/solution interface reaches a critical value corresponding to the critical breakdown potential, the anions adsorbed on the oxide film enter and penetrate the film.	(i) Favored sites for ion migration are suggested to be high-energy regions like grain boundaries and impurities where thinner passive films are produced. (ii) If the aggressive ions meet a metal cation, contaminated film is produced that encourages further ions to penetrate the film. Then, this process continues as an autocatalytic reaction.	(i) Did not explain the observation that pits often form from mechanical breaks in the oxide film or from scratches.

2	Lin et al. [65] 1981	(i) Presented a theoretical model to explain the chemical breakdown of passive film.	(i) Proposed that metal vacancies may accumulate as a result of the diffusion of metal cations from the metal/film to the film/solution interface, forming voids at the metal/film interface. When the voids grow to a critical size the passive film will collapse leading to pit growth.	(i) Surface discontinuities such as grain boundaries and so forth were not considered in developing the model. (ii) No direct observation of void formation was made. (iii) As the measured induction times usually show a large scatter, definite quantitative agreement is difficult to obtain.

Later, Chao et al. [100] proposed a model in which the growth of the passive film was explained by the transport of both anions (e.g., oxygen ion) and cations (e.g., metal ion). Diffusion of anion from film-solution interface to metal-film interface results in thickening of the film. Cation diffusion from the metal-film interface to the film-solution interface results in the creation of metal vacancies at the metal/film interface. These metal vacancies usually "submerge" into the metal itself. However, if the cation diffusion rate is higher than the rate of vacancy submergence into the bulk metal, the metal vacancies will increase leading to the formation of voids at the metal/film interface. This process is known as "pit incubation." Subsequently, when the void reaches a critical size, the pit incubation period ends leading to the local rupture of the passive film. This eventually results in pit growth at that local site. Based on this theory, Chao et al. expressed a criterion for pit "initiation" as stated below.

$$(J_{ca} - J_m) \times (t - \tau) = \xi,$$

(1)

where, J_{ca} is the cation diffusion rate in the film, J_m is the rate of submergence of the metal vacancies into the bulk metal, t is the time required for metal vacancies to accumulate to a critical amount x, τ is a constant.

Also, in this model, the role of the halide ion in accelerating the film breakdown by increasing J_{ca} was suggested.

The ion penetration and migration theories do not include the effect of mechanical breakdown of the oxide film that may result because of the scratches from which pits can nucleate, nor is the mechanical breakdown of the oxide film included those results from strain and local cracking of the oxide film.

In addition Lin et al. [65] have proposed a "point defect" model for anodic films to calculate J_{ca} for "thin" films on the order of 10–40 A. Also, the "point defect" model could be used to calculate incubation times. Although, the "point defect" model was one of the most detailed models proposed, this model has some limitations as mentioned in Table 4.

Mechanical Film Breakdown Theories— Chemicomechanical Breakdown Theories

Pit nucleation models proposed so far based on the concepts of the "chemicomechanical" breakdown of films have not included the effect of externally applied stresses (see Table 7). Sato [63, 64] showed that a significant film pressure always acted on "thin" films that he attributed to "electrostriction." Sato expressed a relation between the film pressure, thickness, and surface tension of the film as follows:

$$p = p_0 + \left[\frac{(\delta(\delta - 1)\xi^2)}{8\pi} \right] - \frac{\gamma}{L},$$

(2)

where p is the film pressure, p_o is the atmospheric pressure, δ is the film dielectric constant, ξ is the electric field, γ is the surface tension, L is the film thickness.

Table 7: Chemicomechanical breakdown theories

	Proposed by	Summary	Description	Limitations
1	Sato 1971 [62]	(i) Proposed a breakdown mechanism for anodic films from thermodynamic considerations.	(i) Showed that thin films always contain film pressure due to "electrostriction." (ii) Hypothesized that both the surface tension of the film and the film thickness have a significant effect on film pressure. (iii) Proposed that adsorption of chloride ions, depending on their concentration, greatly reduces surface tension.	Experimental proof is not found.
2	Sato 1982 [63, 64]	(i) Derived an equation for the work required to form a cylindrical breakthrough pore in the passive film.	(i) Proposed that for a pit nucleus to grow to macroscopic size a critical radius corresponding to critical pore formation energy must be exceeded.	Experimental proof is not found. Microstructural parameters such as grain boundaries, inclusions that may influence pitting "initiation" were not considered.

According to his hypothesis, both γ and L have significant influence on film pressure P. Based on this relation, Sato suggested that the adsorption of chloride ion significantly reduces the surface tension γ thereby increasing. Also, he proposed that when P is above the critical value, the film might break down. In addition, Sato proposed that breakdown of the film occurs when it attains a thickness at which mechanical stresses caused by "electrostriction" become critical. Therefore, building up of critical stresses in the film could cause pitting.

In addition to the aforementioned theory, some researchers have observed the influence of mechanically produced discontinuities (such as scratches in the passive film) on the formation of pits along those scratches [98]. If there is a scratch in the passive film that sets up a local anodic site, which will, eventually, be the preferred site for pit to form, this smaller anode/cathode ratio results in higher local potential leading to the nucleation of pits. Other researchers proposed a similar theory that is related to the value of product of the length of the discontinuity and the current density. Assuming a unidirectionally growing pit, if this value exceeds a critical value, the discontinuity such as "fissures" in the oxide film may form a local area of low pH leading to the formation of pits from them. This happens due to the difference in the pH at the local site (fissure) when compared to the bulk solution. It was proposed that a fissure of size in the order of 10^{-6} cm could be a limiting condition for this to happen [101].

Hoar and Jacob [94] also assumed that the presence of pores or "flaws" could mechanically stress and damage the passive films in contact with an aggressive solution. Moreover, Hoar assumed that aggressive anions would replace water and reduce surface tension at the solution-film interface by repulsive forces between particles, producing cracks.

In conclusion, there is no full agreement among the researchers regarding the mechanisms of pit nucleation. However, as the pitting process itself is a complex one, the commonly accepted view is that the first step in the pit nucleation process is the localized adsorption of aggressive anions on the surface of the passivated metal. Several experimental studies also have indicated that the preferred sites for the passage of anions through the oxide film are the discontinuities present in an alloy. Such discontinuities are nonmetallic inclusions; second-phase precipitates, pores or voids, grain or phase boundaries, and other mechanical damages [69]. These discontinuities eventually may become pit nucleation sites. The aforementioned theories on pit nucleation are based purely on electrochemical concepts. However, the breakdown of surface film is dependent not only on the solution conditions (e.g., pH) and the electrochemical state at the metal/ solution interface, but also on the nature of the material as well as the stress state. In addition, the aforementioned pit nucleation mechanisms did not take into account the material parameters such as the microstructural effects, inherent discontinuities such as voids, inclusions, second-phase particles as well as the externally applied stress. Moreover, localized corrosion also may take place at slip bands during fatigue loading [102].

Once the pit is formed, the rate of pit growth is dependent mainly on the material, local solution conditions, and the state of stress. Cracks have been observed to form from pits under cyclic loading conditions. Therefore, to estimate the total corrosion fatigue life of an alloy, it is of great importance to develop some realistic models to establish the relationship between pit propagation rate and the stress state. Furthermore, pitting corrosion in conjunction with externally applied mechanical stresses, for example, cyclic stresses, has been shown to severely affect the integrity of the oxide film as well as the fatigue life of a metal or an alloy. Therefore, to understand these phenomena, some models based on pitting corrosion fatigue mechanisms have been proposed as discussed below.

Pitting Corrosion Fatigue

Linear Elastic Fracture Mechanics (LEFM) concepts are widely used to characterize the crack growth behavior of materials under cyclic stresses in different environmental conditions. It is important to note that both pitting theory and crack growth theory have been used in model development as follows. Pit growth rate theory proposed by Godard is combined with fatigue crack growth concepts. The time to form/nucleate a Mode I crack from the pit (under cyclic loading) could be modeled using LEFM concepts. Based on this idea, a few models [103–106] were proposed since 1971 (see Table 8). All of the models assume hemispherical geometry for the pit shape, and the corresponding stress intensity relation is used to determine the critical pit depth using the crack growth threshold (ΔK_{th}) that is found empirically. For hemispherical pit geometry, these models provide a reasonable estimate for the total corrosion fatigue life. However, it is well known that corrosion pit morphology varies widely. Thus, this aspect must eventually be dealt with in LEFM models that attempt to deal with pit growth and the ultimate nucleation of crack(s) from pit(s).

Table 8: Pitting corrosion fatigue models background-references [112, 113, 122–217]

	Proposed by	Summary	Description	Advantages/limitations
1	Hoeppner [67] (1972 - current)	(i) Proposed a model to determine critical pit depth to nucleate a Mode I crack under pitting corrosion fatigue conditions. (ii) Combined with the pit growth rate theory as well as the fatigue crack growth curve fit in a corrosive environment, the cycles needed to develop a critical pit size that will form a Mode I fatigue crack can be estimated.	(i) Using a four-parameter Weibull fit, fatigue crack growth threshold (ΔK_{th}) was found from corrosion fatigue experiments for the particular environment, material, frequency, and load spectrum. (ii) The stress intensity relation for surface discontinuity (half penny-shaped crack) was used to simulate hemispherical pit. I.e. $K = 1.1\sigma\sqrt{\pi a/Q}$, where, σ is the applied stress, a is the pit length, and Q is the function of $a/2c$, S_y. (iii) Using the threshold determined empirically, critical pit depth was found from the stress intensity relation mentioned above. (iv) Then, the time to attain the pit depth for the corresponding threshold value was found using $t = (d/c)^n$, where, t is the time, d is the pit depth, and c is a material/environment parameter.	(i) This model provides a reasonable estimate for hemispherical geometry of the pits. (ii) This model is useful to estimate the total corrosion fatigue life with knowledge of the kinetics of pitting corrosion and fatigue crack growth. (iii) This model did not attempt to propose mechanisms of crack nucleation from corrosion pits. (iv) Quantitative studies of pitting corrosion fatigue behavior of materials can be made using this model. (v) This model is valid only for the conditions in which LEFM concepts are applicable. (vi) Material dependent.
2	Lindley et al. [104]	(i) Similar to Hoeppner's model, a method for determining the threshold at which fatigue cracks would grow from the pits was proposed.	(i) Pits were considered as semielliptical-shaped sharp cracks (ii) Used Irwin's stress intensity solution for an elliptical crack in an infinite plate and came up with the relationship to estimate threshold stress intensity values related to fatigue crack nucleation at corrosion pits. i.e. $\Delta K_{th} = \dfrac{\Delta\sigma\sqrt{(\pi a)}[1.13-0.07(a/c)^{1/2}]}{[1+1.47(a/c)^{1.64}]^{1/2}}$, where, $\Delta\sigma$ is the stress range, a is the minor axis, and c is the major axis of a semi-elliptical crack. (iii) From the observed pit geometry that is, for a/c ratio, threshold stress intensity can be calculated. (iv) For the corresponding a/c ratio, critical pit depth can be estimated.	(i) The proposed stress intensity relation can be used in tension-tension loading situations where stress intensity for pits and cracks is similar. (ii) Critical pit depths for cracked specimens can be estimated using the existing threshold stress intensity values. (iii) This model is valid only for the conditions in which LEFM concepts are applicable. (iv) Material dependent.
3	Kawai and Kasai [105]	(i) Proposed a model based on estimation of allowable stresses under corrosion fatigue conditions with emphasis on pitting. (ii) As corrosion is not usually considered in developing S-N fatigue curves, a model for allowable stress intensity threshold involving corrosion fatigue conditions was proposed.	(i) Considered corrosion pit as an elliptical crack. (ii) Based on experimental data generated on stainless steel, new allowable stresses based on allowable stress intensity threshold were proposed. i.e. $\Delta\sigma_{all} = \Delta K_{th}/F\sqrt{\pi h_{max}}$, where ΔK_{th} can be determined from a da/dN versus ΔK plot for a material, h_{max} is the maximum pit depth, and F is a geometric factor.	(i) Using this model, allowable stress in relation to corrosion fatigue threshold as a function of time could be estimated. (ii) Material dependent. (iii) This model is valid only for the conditions in which LEFM concepts are applicable.
4	Kondo [111]	(i) Corrosion fatigue life of a material could be determined by estimating the critical pit condition using stress intensity factor relation as well as the pit growth rate relation.	(i) Pit diameter was measured intermittently during corrosion fatigue tests. (ii) From test results, corrosion pit growth law was expressed as $2a = C_p t^{1/3}$, where $2c$ is the pit diameter, t is the time, and C_p is an environment/material parameter. Then, critical pit condition (ΔK_p) in terms of stress intensity factor was proposed by assuming pit as a crack. $\Delta K_p = 2.24\sigma_a\sqrt{\pi(a/Q)}$, where σ_a is the stress amplitude, a is the aspect ratio, and Q is the shape factor. (iii) Critical pit condition was determined by the relationship between the pit growth rate theory and fatigue crack growth rates: $c = c_p(N1f)^{1/3}$, where N is the number of stress cycles, f is the frequency, and $2c$ is the pit diameter. (iv) The pit growth rate dc/dN was developed using ΔK relation as given below $dc/dN = (1/3)C_p^3 f^{-1}\pi^2\pi^2 Q^{-2}(2.24\sigma_a)^4\Delta K^{-4}$ dc/dN was determined using experimental parameter Cp. (v) Finally, the critical pit size $2C_{cr}$ was calculated from the stress intensity factor relation. i.e., $2c_{cr} = (2Q/\pi a)(\Delta K_p/2.24\sigma_a)^2$.	(i) The aspect ratio was assumed as constant. (ii) Material and environment dependent.

As mentioned before, the combined effect of corrosion and the applied cyclic loading have been shown to produce cracks from corrosion pits. In addition, pits have frequently been the source of cracks on aircraft that are operating in fleets. Depending upon the fatigue loading and corrosion conditions, some studies have shown that the crack formation/nucleation site may change from slip bands to corrosion pits [107]. This observation was made when fatigue was tested at reduced strain rates in Al-Li-Cu alloy. Another study also showed an anodic dissolution in slip bands in Al-Li-Zr alloy at high stress levels whereas at low stress levels fatigue cracks nucleated from corrosion pits [108]. Therefore, it was hypothesized that at higher stress levels, conditions are favorable to form cracks from slip bands before the corrosion pit reaches the critical condition to favor the nucleation of crack from it. In addition, a recent study also showed that larger pit was formed at lower stress and frequency. It also was observed in 2024-T3 (bare) aluminum alloy in NaCl solution that once pits formed from the constituent particles, because of the applied cyclic stresses, the pits coalesced laterally and in depth to form larger pits from which crack was observed to nucleate [109]. Therefore, modeling the transition of a pit first to a "short" crack and then to a "long" crack is considered to be important in characterizing the total corrosion fatigue life of a material as discussed in the next section [66, 106, 110].

ENVIRONMENTAL EFFECTS ON "SHORT" CRACK BEHAVIOR OF MATERIALS

A few "small" crack studies under corrosion fatigue conditions have been performed to characterize the transition of a pit to a "small" crack. In 2024 aluminum alloy, Piascik and Willard have shown a three times increase in crack growth rates of "small" cracks in salt water environment when compared to air. Moreover, their studies clearly have observed the transition of pits formed at the constituent particles to intergranular "microcracks" and then to transgranular fracture path once the crack reaches the depth of 100 mm. In addition, the increase in "small" crack growth rates was observed even at very low mode IΔK (<1 MPa $\sqrt{m}$). As well, Kondo [111] also observed in two low alloy steels that "short" cracks from pits propagated at ΔK that is well below the threshold value of a long crack for these materials.

In a recent in situ fatigue study, prior pitted 2024-T351 and 7075-T651 aluminum alloy specimens exhibited faster crack growth rates in the "short" crack regime when compared to specimens without prior corrosion damage (Hoeppner [now Taylor] [112]). This study showed that prior corrosion damage did influence the "small" crack growth rates. It also was observed that the 7075 aluminum alloy specimen had faster crack growth rates compared to the 2024 aluminum alloy specimens. Also, in this study cracks were observed to form from pits on the prior corroded specimens whereas on the specimens without any prior corrosion damage, cracks formed from constituent particles.

In addition to a few previous studies (Hoeppner, 1979, [103]) in which pitting was modeled statistically with different materials and specimen types, recently, as discussed before in this paper, there was a study demonstrating that corrosion fatigue induced "short" crack formation from pits [113]. Also, recent studies [73,110, 112] have shown that pits form in different shapes depending upon environment and loading conditions in contradiction to general assumption that pits have hemispherical shape. Although this assumption simplifies the modeling part of research [111], further studies to characterize the formation of cracks from pits in the "short" crack regime must be evaluated as indicated by A. Hoeppner [112]. Apart from these studies the literature search has not found any "short" crack studies to evaluate the formation of cracks from pits and their crack morphologies and paths. Moreover, fretting mechanism(s) in conjunction with fatigue and corrosion may further aggravate this.

CONCLUSIONS AND RECOMMENDATIONS

The review of the literature clearly shows that much progress has been made on modeling the effects of corrosion on material behavior and structural integrity. It is clear that to date the models have centered on characterizing the corrosion and modeling the effects of the corrosion as one or more of the following:

- section change that affects the area/volume that modifies the stress,
- formation/nucleation of localized debris that may modify the stress (part of pillowing) that modifies the stress or stress intensity;
- nucleation of intergranular corrosion that is involved in pillowing that modifies the stress or stress intensity;
- nucleation of localized corrosion (pitting, fretting, etc.) that modifies the local stress and may ultimately nucleate cracks;
- production of products of corrosion that produce localized embrittlement effects that may alter the material behavior and produce accelerated crack propagation.

All of the above have been reviewed in the preceding sections and lead to the recognition that one of the most pressing issues to be resolved is the actual quantitative characterization of the corrosion in relation to the physical damage state that is underway. Some of this has been accomplished in the past with the efforts of the past at the University of Utah as discussed in the earlier sections of this paper. From the work of L. Grimes at Utah as well as additional efforts at the University of Utah, the use of the confocal microscope will be of great assistance in characterizing the three-dimensional (3D) surface "damage" that results from corrosion of various forms.

Within the last few years interest in corrosion and the effects of corrosion has picked up in part due to numerous failures in many industries including nuclear power plants, gas and oil pipelines, and aircraft to name a few. Roberge [114–116] has introduced excellent reference books on aspects of corrosion and also a web page (http://www.corrosion-doctors.org/) that contain a wealth of information related to many of the topics covered in this paper. A recent issue of business week [117] states that the USA DOD spends "22.9 billion a year fighting rust". There is little doubt that this number will become much larger and more of the structures in use in aircraft and many other applications age and it is unlikely that more funds will be appropriated to replace many aging aircraft components. Thus, many of the issues covered herein will become more important in both the design, operational, and maintenance strategies to combat the issue of corrosion. This also is clear from the fact that the USA DOD has established a Corrosion Policy and Oversight Office Congress in the pentagon as was mandated by the US Congress in 2003. It remains to be seen whether this will result in significant cost savings to combat corrosion and reduce the number of accidents from corrosion-related issues.

Even though fracture mechanics-based modeling has been extremely useful in modeling the effects of corrosion on structural integrity it has taken many simplifications and, depending on the manner in which the fracture mechanics is used in the model, has resulted in downgrading the real corrosion characterization issue and understanding the 3D nature of the corrosion degradation process. New tools and models will have to be brought to bear on the formation/nucleation and growth of the corrosion with or without load of either sustained (SCC) or cyclic nature (EANC/F) (Environmentally assisted nucleation and cracking with fatigue loading). Furthermore the transitions of corrosion to actual cracks will have to be understood to improve the models that currently exist and any new ones that may be developed. Aspects of this were discussed by Hoeppner [118] and Swift [119] in recent ICAF meetings. No doubt more attention will be focused on this in the future.

The characterization of chemically dependent short crack propagation and modeling of it will have to be much better understood. One area not addressed in the article is the effect of either prior corrosion and/or concomitant corrosion on either fatigue crack propagation or stress-corrosion cracking. Both of these issues are extremely important to the overall area of model development and consideration should be given to expanding at many laboratories in the future.

The importance of corrosion to DOD activities within the USA has recently be noted

ACKNOWLEDGMENTS

The authors wish to express appreciation to the former Dr. John DeLuccia who provided valuable comment and insights for the preparation of this paper. In addition, they express our deep appreciation to FASIDE International Inc. and the University of Utah for provision of facilities and a magnificent library system. Ms. Amy Taylor and Dr. Chandrasekaran provided much valuable input over a period of years. They appreciate all their efforts. The following three individuals provided extensive discussions on the topics covered herein: Mr. Nick Bellinger and Mr. Jerzy Komorowski (both of NRC-IAR-Canada), and Mr. Craig Brooks

(AP/ES-USA). The authors are grateful for the many discussions and interactions with them over many years.

REFERENCES

1.   "Stress corrosion cracking in aircraft structural materials," in Proceedings of the Symposium held by the Structures and Materials Panel of AGARD, vol. 18 of AGARD Conference Proceedings Series, NATO-AGARD, Turin, Italy, April, 1967.

2.   "Fundamental aspects of stress corrosion cracking," in Proceedings of the conference held at the Ohio State University, R. W Staehle, A. J. Forty, and D. van Rooyen, Eds., National Association of Corrosion Engineers, Ohio, USA, September, 1967.

3.   "Effects of environment and complex load history on fatigue life, ASTM STP 462," in Proceedings of the Symposium on Effects of Environment and Complex Load History on Fatigue Life, M. Rosenfeld, D. W. Hoeppner, and R. I. Stephens, Eds., ASTM, Atlanta, Ga, USA, September, 1986.

4.   "Corrosion fatigue: chemistry, mechanics, and microstructure," in Proceedings of the Conference held at the University of Connecticut, O. Devereux, A. J. McEvily, and R. W. Staehle, Eds., National Association of Corrosion Engineers, Connecticut, Conn, USA, June, 1971.

5.   B. F. Brown, Stress-Corrosion Cracking in High Strength Steels and in Titanium and Aluminum Alloys, Naval Research Laboratory, Washington, DC, USA, 1972.

6.   L. R. Hall, R. W. Finger, and W. F. Spurr, "Corrosion fatigue crack growth in aircraft structural materials," Tech. Rep. number AFML-TR-73-204, Boeing Company, 1973.

7.   D. Pettit, J. Ryder, W. Krupp, and D. Hoeppner, "Investigation of the effects of stress and chemical environments on the prediction of fracture in aircraft structural materials," Tech. Rep. number AFML-TR-74-183, Lockheed California Company, 1974.

8.   "Corrosion-fatigue technology, ASTM STP 642," in Proceedings of the Symposium held in Denver, CO, H. L. Craig Jr., T. W. Crooker, and D. W. Hoeppner, Eds., ASTM, Denver, Colo, USA, November, 1976.

9. "Aircraft corrosion," in Proceedings of the 52nd Meeting of the AGARD Structures and Materials Panel, no. 315, NATO-AGARD, Cesme, Turkey, April, 1981.

10. "Corrosion fatigue," in Proceedings of the 52nd Meeting of the AGARD Structures and Materials Panel, no. 316, NATO-AGARD, Cesme, Turkey, April, 1981.

11. "Corrosion fatigue, STM STP 801," in Proceedings of the Symposium on Corrosion Fatigue: Mechanics, Metallurgy, Electrochemistry, and Engineering, T. W. Crooker and B. N. Leis, Eds., ASTM, St. Louis, Mo, USA, October, 1981.

12. D. W. Hoeppner and V. Chandrasekaran, "Corrosion and corrosion fatigue predictive modeling-state of the art review," Tech. Rep., FASIDE International, 1998.

13. W. Wallace, D. Hoeppner, and P. V. Kandachar, "Aircraft corrosion: causes and case histories," inAGARD Corrosion Handbook, vol. 1, North Atlantic Treaty Organization, 1985.

14. ASM Handbook, vol. 13, ASM International, Metals Park, Ohio, USA, 1987.

15. The Boeing Company, "Corrosion prevention and control, manual for training operators of Boeing commercial," Aircraft, Seattle, Wash, USA, 1988.

16. NTSB Metallurgist's, "Aloha airlines flight 243," Tech. Rep. number 88–85, Materials Laboratory, 1988.

17. "Environment induced cracking of metals," in Proceedings of the International Conferencen, R. P. Gangloff and M. B. Ives, Eds., National Association of Corrosion Engineers NACE-10, Kohler, Wash, USA, October, 1988.

18. ASM Handbook, vol. 18, Friction Lubrication and Wear Technology ASM International, Metals Park, Ohio, USA, 1992.

19. "Naval Aviation Safety Program," US Navy, OPNAVINST 3750.6Q CH-1, OP-05F, 1991.

20. Federal Aviation Administration, "Aircraft accident and incident synopses related to corrosion, fretting, and fatigue for the period 1976–1993," 1994.

21. National Transportation Safety Board, "Aircraft accident and incident synopses related to corrosion, fretting, and fatigue for the period 1975–1993," 1994.

22. F. Karpala and O. L. Hageniers, "Characterization of corrosion and development of a broadboard model of a D sight aircraft inspection system," Report to DOT phase 1, Diffracto, 1994.

23. United States Air Force, "Navy, and army, aircraft accident and incident synopses related to corrosion, fretting, and fatigue," 1994.

24. G. Cooke, P. J. Vore, C. Gumienny, and G. Cooke Jr., "A study to determine the annual direct cost of corrosion maintenance for weapon systems and equipment in the United States Air Force," Final report contract number F09603-89-C-3016, SEPT., 1990.

25. W. Schütz, "Corrosion fatigue-the forgotten factor in assessing durability," in Proceedings of the 18th Symposium on the International Committee of Aeronautical Fatigue, (ICAF '95), J. M. Grandage and G. S. Jost, Eds., vol. 1 of Estimation, Enhancement and Control of Aircraft Fatigue Performance, pp. 1–52, EMAS, Melbourne, Australia, May, 1995.

26. S. J. Swift, "The aero commander chronicle," in Proceedings of the 18th Symposium on the International Committee of Aeronautical Fatigue, (ICAF '95), J. M. Grandage and G. S. Jost, Eds., vol. 1 of Estimation, Enhancement and Control of Aircraft Fatigue Performance, pp. 507–530, EMAS, Melbourne, Australia, May, 1995.

27. D. W. Hoeppner, L. Grimes, A. Hoeppner, J. Ledesma, T. Mills, and A. Shah, "Corrosion and fretting as critical aviation safety issues," in Proceedings of the 18th Symposium on the International Committee of Aeronautical Fatigue, (ICAF '95), J.M. Grandage and G.S. Jost, Eds., vol. 1 of Estimation, Enhancement and Control of Aircraft Fatigue Performance, pp. 87–106, EMAS, Melbourne, Australia, May, 1995.

28. C. L. Brooks, K. Liu, and R.G. Eastin, "Understanding fatigue failure analyses under random loading using a C-17 test report," in Proceedings of the 18th Symposium on the International Committee of Aeronautical Fatigue, (ICAF '95), J. M. Grandage and G. S. Jost, Eds., vol. 1 of Estimation, Enhancement and Control of Aircraft Fatigue Performance, pp. 449–468, EMAS, Melbourne, Australia, May, 1995.

29. ASM Handbook, vol. 19, Fatigue and Fracture ASM International, Metals Park, Ohio, USA, 1996.

30.   C. G. Schmidt, J. E. Crocker, J. H. Giovanola, C. H. Kanazawa, and Schockey, "Characterization of early stages of corrosion fatigue in aircraft skin," Final report, contract No. 93-G-065, SRI International Final Report to DOT, Menlo Park, Calif, USA, Report No. DOT/FAA/AR-95/108, 1996.

31.   G. K. Cole, G. Clark, and P. K. Sharp, "Implications of corrosion with respect to aircraft structural integrity," Tech. Rep. number DSTO-RR-0102, AMRL, Melbourne, Australia, 1997.

32.   Aging of U.S. Air Force Aircraft, Report of the committee on aging of U.S. Air Force Aircraft, NMAB (National Materials Advisory Board), Commission on Engineering and Technical Systems, National Research Council, Publication NMAB-488-2, 1997.

33.   Boeing, "Corrosion damage assessment framework, corrosion/fatigue effects on structural integrity," Tech. Rep. number D500-13008-1, USAF contract No. F9603-97-C-0349, 1998.

34.   C. Paul and T. Mills, "Corrosion/fatigue," in Proceedings of the Aerospace Materials Conference, 1998.

35.   "A study to determine the cost of corrosion maintenance for weapon systems and equipment in the United States Air Force," Final Report Contract number F09603-95-D-0053, 1998.

36.   "Fatigue in the presence of corrosion," in Proceedings of the Workshop of the RTO Applied Technology (AVT) Panel, vol. 18 of RTO Proceedings, no. AC/323(AVT) TP/8, NATO, Research and Technology Organization, Corfu, Greece, October, 1998.

37.   D. Jones, Principles and Prevention of Corrosion, Macmillan Publishing, New York, NY, USA, 1992.

38.   D. W. Hoeppner, "Estimation of component life by application of fatigue crack growth threshold knowledge," in Fatigue, Creep and Pressure Vessels for Elevated Temperature Service, C. W. Lawton and R. R. Seeley, Eds., pp. 1–83, The American Society of Mechanical Engineers, New York, NY, USA, 1981.

39.   D. W. Hoeppner, "Parameters that input to application of damage tolerance concepts to critical engine components," in Proceedings of the AGARD Conference Damage Tolerance Concepts for Critical Engine Components, no. AGARD-CP 393, pp. 4.1–4.16, NATO-AGARD, San Antonio, Tex, USA, April, 1985.

40. American Society for Metals, Metals Handbook, vol. 13 of Corrosion, American Society for Metals (ASM), Metals Park, Ohio, USA, 9th edition, 1987.

41. A. Alvarez, "Corrosion on aircraft in marine-tropical environments: a technical analysis," Material Performance, vol. 36, no. 5, pp. 33–38, 1997.

42. B. M. Suyitno and T. Sutarmadji, "Corrosion control assessment for Indonesian aging aircraft," Anti-corrosion Methods and Materials, vol. 44, no. 2, pp. 115–122, 1997.

43. D. J. Groner, "US Air Force aging aircraft corrosion," in Current Awareness Bulletin, Structures Division, Wright Laboratory, Spring, 1997.

44. R. S. Piascik, R. G. Kelly, M. E. Inman, and S. A. Willard, "Fuselage lap splice corrosion," Tech. Rep. number WL-TR-96-4094, ASIP, 1996.

45. J. F. McIntyre and T. S. Dow, "Intergranular corrosion behavior of aluminum alloys exposed to artificial seawater in the presence of nitrate anion," Corrosion, vol. 48, no. 4, pp. 309–319, 1992.

46. J. J. Thompson, E. S. Tankins, and V. S. Agarwala, "A heat treatment for reducing corrosion and stress corrosion cracking susceptibilities in 7xxx aluminum alloys," Materials Performance, pp. 45–52, 1987.

47. J. V. Rinnovatore, K. F. Lukens, and J. D. Corrie, "Exfoliation corrosion of 7075 aluminum die forgings,"Corrosion, vol. 29, no. 9, pp. 364–372, 1973.

48. T. B. Mills, The combined effects of prior-corrosion and aggressive chemical environments on fatigue crack growth behavior in aluminum alloy 7075—T651, Ph.D. dissertation, University of Utah, 1997.

49. T. B. Mills, The effects of exfoliation corrosion on the fatigue response of 7075-T651 aluminum plate, M.S. thesis, University of Utah, 1994.

50. J. P. Chubb, T. A. Morad, B. S. Hockenhull, and J.W. Bristow, "The effect of exfoliation corrosion on the fatigue behavior of structural aluminum alloys," in Structural Integrity of Aging Airplanes, S. N. Atluri, S. G. Sampath, and P. Tong, Eds., pp. 87–97, Springer, New York, NY, USA, 1991.

51. D. D. Swartz, M. Miller, and D. W. Hoeppner, "Chemical environments in commercial transport aircraft and their effect on corrosion fatigue crack propagation," in Proceedings of the 14th symposium of the International Committee on Aeronautical Fatigue, J.M. Grandage and G. S. Jost, Eds., vol. 1 ofEstimation, Enhancement and Control of Aircraft Fatigue Performance, pp. 353–364, Melbourne, Australia, May, 1995.

52. J. Kramer and D. W. Hoeppner, "Effects of cyclic immersion in 3.5% NaCl solution on fatigue crack propagation rates in aluminum 2024-T351," in Proceedings of the USAF Structural Integrity Program Conference, vol. 2, no. WL-TR-96-4093, pp. 1089–1112, 1995.

53. T. B. Mills, D. J. Magda, S. E. Kinyon, and D. W. Hoeppner, "Fatigue crack growth and residual strength analyses of service corroded 2024-T3 aluminum fuselage panels," Tech. Rep., Oklahoma City Air Logistics Center and Boeing Defense and Space Group, University of Utah, 1995.

54. J. P. Komorowski, N. C. Bellinger, and R. W. Gould, "The role of corrosion pillowing in NDI and in the structural integrity of fuselage joints," in Proceedings of the 19th Symposium of the Industrial College of the Armed Forces, (ICAF '97), Fatigue in New and Aging Aircraft, 1997.

55. N. C. Bellinger, J. P. Komorowski, and R. W. Gould, "Damage tolerance implications of corrosion pillowing on fuselage lap joints," American Institute of Aeronautics and Astronautics Journal, vol. 3, pp. 317–320, 1997.

56. M. Jayalakshmi and Muralidharan Muralidharan, "Empirical and deterministic models of pitting corrosion—an overview," Corrosion Reviews, vol. 14, no. 3-4, pp. 375–402, 1996.

57. V. R. Evans, "XVI.—the passivity of metals. Part II. The breakdown of the protective film and the origin of corrosion currents," Journal of the Chemical Society, pp. 92–110, 1929.

58. T. P. Hoar, "The production and breakdown of the passivity of metals," Corrosion Science, vol. 7, pp. 341–355, 1967.

59. T. P. Hoar and G. C. Wood, "The sealing of porous anodic oxide films on aluminum," Electrochemica Acta, vol. 7, pp. 333–353, 1962.

60. H. Bohni and H. H. Uhlig, "Environmental factors affecting the critical pitting potential of aluminum,"Journal of the Electrochemical Society, vol. 116, pp. 906–910, 1969.

61. Ja. M. Kolotyrkin and M. Ya, "Effects of anions on the dissolution kinetics of metals," Journal of the Electrochemical Society, vol. 108, no. 3, pp. 209–216, 1961.

62. N. Sato, "A theory for breakdown of anodic oxide films on metals," Electrochemica Acta, vol. 16, no. 10, pp. 1683–1692, 1971.

63. N. Sato, "Anodic breakdown of passive films on metals," Journal of the Electrochemical Society, vol. 129, no. 2, pp. 255–260, 1982.

64. N. Sato, "The stability of pitting dissolution of metals in aqueous solution," Journal of the Electrochemical Society, vol. 129, no. 2, pp. 260–264, 1982.

65. L. F. Lin, C. Y. Chao, and D. D. Macdonald, "A point defect model for anodic passive films II. Chemical breakdown and pit initiation," Journal of the Electrochemical Society, vol. 128, no. 6, pp. 1194–1198, 1981.

66. D. W. Hoeppner, D. Mann, and J. Weekes, "Fracture mechanics based modeling of corrosion fatigue process," in Proceedings of the 52nd Meeting of the AGARD Structural and Materials Panel, Corrosion Fatigue, Cesme, Turkey, April, 1981.

67. D. W. Hoeppner, "Corrosion fatigue considerations in materials selections and engineering design," inCorrosion Fatigue: Chemistry, Mechanics, and Microstructure, pp. 3–11, NACE, 1972.

68. D. Jones, Principles and Prevention of Corrosion, Macmillan Publishing, New York, NY, USA, 1992.

69. S. Szklarska, Pitting Corrosion of Metals, National Association of Corrosion Engineers (NACE), Houston, Tex, USA, 1986.

70. J. E. Hatch, Ed., Aluminum Properties and Physical Metallurgy, American Society for Metals (ASM), Metals Park, Ohio, USA, 1984.

71. D. J. McAdam and G. W. Gell, "Pitting and its effect on the fatigue limit of steels corroded under various conditions," Journal of the Proceedings of the American Society for Testing Materials, vol. 41, pp. 696–732, 1928.

72.   R. T. Foley, "Localized corrosion of aluminum alloys—a review," Corrosion, vol. 428, no. 5, pp. 277–288, 1986.

73.   L. Grimes, A comparative study of corrosion pit morphology in 7075-T6 aluminum alloy, M.S. thesis, University of Utah, 1996.

74.   Liversidge, "On the corrosion of aluminum," in Chemical News, LXXI, 1895.

75.   G. D. Bengough and O. F. Hudson, "Aluminum, fourth report to the corrosion committee," Journal of the Institute of Metals, vol. 21, no. 1, p. 105, 1919.

76.   H. T. Barnes and G . W. Shearer, "A hydrogen peroxide cell," Journal of Physical Chemistry, vol. 12, p. 155, 1908.

77.   W. R. Dunstan and J. R. Hill, "The passivity of iron and certain other metals," Journal of the Chemical Society, Transactions, vol. 99, pp. 1853–1866, 1911. ·

78.   H. P. Godard, The Corrosion of Light Metals, John Wiley and Sons, New York, NY, USA, 1967.

79.   R. K. Hart, "The oxidation of aluminum in dry and humid oxygen atmospheres," The Proceedings of the Royal Society, vol. A236, p. 68, 1956.

80.   G. E. Thompson, K. Shimizu, and G. C. Wood, "Observation of flaws in anodic films on aluminum,"Nature, vol. 286, pp. 471–472, 1980.

81.   R. Seligman and P. Williams, "The action on aluminum of hard industrial waters," Journal of the Institute of Metals, vol. 23, pp. 159–184, 1920.

82.   R. S. Alwitt, "The growth of hydrous oxide films on aluminum," Journal of the Electrochemical Society, vol. 121, no. 10, pp. 1322–1328, 1974.

83.   T. E. Graedel, "Corrosion mechanisms for aluminum exposed to the atmosphere," Journal of the Electrochemical Society, vol. 136, no. 4, p. 204C, 1989.

84.   I. L. Rosenfield and I. S. Danilov, "Electrochemical aspects of pitting corrosion," Corrosion Science, vol. 7, p. 129, 1967.

85.   A. Broli and H. Holtan, "Use of potentiokinetic methods for the determination of characteristic potentials for pitting corrosion of aluminum in a deaerated solution of 3%NaCl," Corrosion Science, vol. 13, no. 4, pp. 237–246, 1973.

86. H. P. Godard, The Corrosion of Light Metals, John Wiley and Sons, New York, NY, USA, 1967.

87. S. Dallek and R. T. Foley, "Propagation of pitting on aluminum alloys," Journal of the Electrochemical Society, vol. 125, no. 5, pp. 731–733, 1978.

88. T. H. Nguyen and R. T. Foley, "On the mechanism of pitting of aluminum," Journal of the Electrochemical Society, vol. 126, no. 11, pp. 1855–1860, 1979.

89. M. Baumgartner and H. Kaesche, "Aluminum pitting in chloride solutions: morphology and pit growth kinetics," Corrosion Science, vol. 31, pp. 231–236, 1990.

90. L. Ma, Pitting effects on the corrosion fatigue life of 7075-T6 aluminum alloy, dissertation, University of Utah, 1994.

91. G. S. Chen, C. Liao, K. Wan, M. Gao, and R. P. Wei, "Pitting corrosion and fatigue crack nucleation," inEffects of the Environment on the Initiation of Crack Growth, ASTM STP 1298, W. A. Van Der Sluys, R. S. Piascik, and R. Zawierucha, Eds., pp. 18–33, American Society for Testing and Materials, 1997.

92. H. H. Uhlig, "Adsorbed and reaction-product films on metals," Journal of the Electrochemical Society, vol. 97, no. 11, pp. 215c–220c, 1950.

93. H. P. Leckie and H. H. Uhlig, "Environmental factors affecting the critical potential for pitting in 18-8 stainless steel," Journal of the Electrochemical Society, vol. 113, no. 2, pp. 1262–1267, 1966.

94. T. P. Hoar and W. R. Jacob, "Breakdown of passivity of stainless steel by halide ions," Nature, vol. 216, pp. 1299–1301, 1967.

95. Ya, M. Kolotyrkin, "Effects of anions on the dissolution kinetics of metals," Journal of the Electrochemical Society, vol. 108, no. 3, pp. 209–216, 1961.

96. H. H. Strehblow, B. Titze, and B.P. Loechel, "The breakdown of passivity of iron and nickel by flouride,"Corrosion Science, vol. 19, pp. 1047–1057, 1979.

97. H. Bohni, "Localized corrosion," in Corrosion Mechanisms, F. Mansfeld, Ed., pp. 285–328, Marcel Dekker, New York, NY, USA, 1987.

98. J. A. Richardson and G. C. Wood, "A study of the pitting corrosion of Al by scanning electron microscopy," Corrosion Science, vol. 10, no. 5, pp. 313–323, 1970.

99. N. Nilsen and E. Bardal, "Short duration tests and a new criterion for characterization of pitting resistance of Al alloys," Corrosion Science, vol. 17, pp. 635–646, 1977.

100. C. Y. Chao, L. F. Lin, and D. D. Macdonald, "A point defect model for anodic passive films I. film growth kinetics," Journal of the Electrochemical Society, vol. 128, no. 6, pp. 1187–1194, 1981.

101. J. R. Galvele, "Transport processes and the mechanism of pitting of metals," Journal of the Electrochemical Society, vol. 123, no. 4, pp. 464–474, 1976.

102. R. Akid, "The role of stress-assisted localized corrosion in the development of short fatigue cracks," inEffects of the Environment on the Initiation of Crack Growth, ASTM STP 1298, W. A. Van Der Sluys, R. S. Piascik, and R. Zawierucha, Eds., pp. 3–17, American Society for Testing and Materials, 1997.

103. D. W. Hoeppner, "Model for prediction of fatigue lives based upon a pitting corrosion fatigue process," in Proceedings of the ASTM-NBS-NSF Symposium, J. T. Fong, Ed., Fatigue Mechanisms, no. ASTM STP 675, pp. 841–870, American Society for Testing and Materials, 1979.

104. T. C. Lindley, P. McIntyre, and P. J. Trant, "Fatigue crack initiation at corrosion pits," Metals Technology, vol. 9, pp. 135–142, 1982.

105. S. Kawai and K. Kasai, "Considerations of allowable stress of corrosion fatigue (focused on the influence of pitting)," Fatigue Fracture of Engineering Materials Structure, vol. 8, no. 2, pp. 115–127, 1985.

106. D. W. Hoeppner, "Corrosion fatigue considerations in materials selections and engineering design," inCorrosion Fatigue: Chemistry, Mechanics, and Microstructure, O. Devereux, A. J. McEvily, and R. W. Staehle, Eds., pp. 3–11, NACE-2, National Association of Corrosion Engineers, 1972.

107. M. Rebiere and T. Magnin, "Corrosion fatigue mechanisms of an 8090 Al Li Cu alloy," Materials Science and Engineering: A, vol. 128, no. 1, pp. 99–106, 1990.

108. G. S. Chen and D. J. Duquette, "Corrosion fatigue of a precipitation-hardened Al-Li-Zr Alloy in a 0.5 M sodium chloride solution," Metallurgical Transactions, vol. 23, no. 5, pp. 1563–1572, 1992.

109. G. S. Chen, M. Gao, and R. P. Wei, "Microconstituent-induced pitting corrosion in aluminum,"Corrosion, vol. 52, pp. 8–15, 1996.

110. L. Ma and D. W. Hoeppner, "The effects of pitting on fatigue crack nucleation in 7075-T6 aluminum alloy," in Proceedings of the FAA/NASA International Symposium on Advanced Structural Integrity Methods for Airframe Durability and Damage Tolerance, vol. 3274, part 1, pp. 425–440, NASA Conference Publication, 1994.

111. Y. Kondo, "Prediction of fatigue crack initiation life based on pit growth," Corrosion Science, vol. 45, no. 1, pp. 7–11, 1989.

112. A. M. Hoeppner, The effect of prior corrosion damage on the short crack growth rates of two aluminum alloys, M.S. thesis, University of Utah, 1996.

113. R. Akid and G. Murtaza, "Environment assisted short crack growth behavior of a high strength steel," inShort Fatigue Cracks, K.J. Miller and E.R. de los Rios, Eds., pp. 193–208, Mechanical Engineering Publications, 1992.

114. P. R. Roberge, Handbook of Corrosion Engineering, McGraw Hill Book, New York, NY, USA, 1999.

115. P. R. Roberge, Corrosion Engineering: Principles and Practice, McGraw Hill Book, New York, NY, USA, 2008.

116. P. R. Roberge, Corrosion Inspection and Monitoring, Wiley-Interscience, Hoboken, NJ, USA, 2007.

117. Bloomberg Business Week, page 37, June 12, 2011.

118. D. W. Hoeppner, "A review of corrosion fatigue and corrosion/fatigue considerations in aircraft structural design," in Proceedings of the 22nd Symposium of the International Committee of Aeronautical Fatigue, M. Guillaume, Ed., vol. 1 of ICAF 2003-Fatigue of Aeronautical Structures as an Engineering Challenge, pp. 425–438, EMAS Publishing, Lucerne, Switzerland, May, 2003.

119. S. Swift, "Rusty diamond," in Proceedings of the 24th Symposium of the International Committee on Aeronautical Fatigue, L.

Lazzeri and A. Salvetti, Eds., ICAF 2007 Durability and Damage Tolerance of Aircraft Structures: Metals vs. Composites, Naples, Italy, May, 2007.

120. ASTM Standards volume 03.02-Wear and Erosion; Metal Corrosion, ASTM, Philadelphia, Pa, USA, 1994.

121. ASTM Standards volume 03.01-Metals-Mechanical Testing; Elevated and Low Temperature Tests; Metallography, ASTM, Philadelphia, Pa, USA, 1998.

122. "Behavior of short cracks in airframe components," in Proceedings of the 55th Meeting of the AGARD Structures and Materials Panel, no. 328, Toronto, Canada, September, 1983.

123. R. K. Bolinbroke and J. E. King, "The growth of short fatigue cracks in titanium alloys IMI550 and IMI318," in Small Fatigue Cracks, R. O. Ritchie and J. Lankford, Eds., pp. 129–144, Metallurgical Society, 1986.

124. A. Boukerrou and R.A. Cottis, "The influence of corrosion on the growth of short fatigue cracks in structural steels," in Short Fatigue Cracks, K. J. Miller and E. R. de los Rios, Eds., pp. 209–217, Mechanical Engineering Publications, 1992.

125. J. L. Breat, F. Mudry, and A. Pineau, "Short crack propagation and closure effects in A508 steel," Fatigue Fracture of Engineering Materials Structures, vol. 6, pp. 349–358, 1983.

126. C. W. Brown, J. E. King, and M. A. Hicks, "Effects of microstructure on long and short crack growth in nickel base super alloys," Metal Science, vol. 18, pp. 374–380, 1984.

127. C. W. Brown and D. Taylor, "The effects of texture and grain size on the short fatigue crack growth rates in Ti-6Al-4V," in Fatigue Crack Threshold Concepts, D.L. Davidson and S. Suresh, Eds., pp. 433–446, AIME, 1984.

128. D. W. Cameron, Perspectives and insights on the cyclic response of metal, Ph.D. dissertation, University of Toronto, 1984.

129. K. S. Chan and J. Lankford, "The role of microstructural dissimilitude in fatigue and fracture of small cracks," Acta Metallurgica, vol. 36, pp. 193–206, 1988.

130. P. Clement, J. P. Angeli, and A. Pineau, "Short crack behavior in nodular cast iron," Fatigue Fracture of Engineering Materials and Structures, vol. 7, pp. 251–265, 1984.

131. D. L. Davidson, "Small and large fatigue cracks in aluminum alloys," Acta Metallurgica, vol. 36, no. 8, pp. 2275–2282, 1988.

132. D. L. Davidson, J. B. Campbell, and R. A. Page, "The initiation and growth of fatigue cracks in a titanium aluminide alloy," Metallurgical Transactions: A, vol. 22, pp. 377–391, 1991.

133. E. R. De Los Rios, Z. Tang, and K. J. Miller, "Short crack fatigue behavior in a medium carbon steel,"Fatigue Fracture of Engineering Materials and Structures, vol. 7, pp. 97–108, 1984.

134. E. R. De Los Rios, H. J. Mohamed, and K. J. Miller, "A micromechanic analysis for short fatigue crack growth," Fatigue Fracture of Engineering Materials and Structures, vol. 8, pp. 49–63, 1985.

135. E. R. De Los Rios, A. Navarro, and K. Hussain, "Microstructural variations in short fatigue crack propagation of a C-Mn steel," in Short Fatigue Cracks, K. J. Miller and E. R. de los Rios, Eds., pp. 115–132, Mechanical Engineering Publications, 1992.

136. V. B. Dutta, S. Suresh, and R. O. Ritchie, "Fatigue crack propagation in dual-phase steels: effects of ferritic-martensitic microstructures on crack path morphology," Metallurgical Transactions: A, vol. 15, pp. 1193–1207, 1984.

137. J. N. Eastabrook, "A dislocation model for the rate of initial growth of stage I fatigue cracks,"International Journal of Fracture, vol. 24, no. 1, pp. 43–49, 1984. ·

138. M. H. El Haddad, N.E. Dowling, T. H. Topper, and K. N. Smith, "J-integral applications for short fatigue crack at notches," International Journal of Fracture, vol. 16, pp. 15–30, 1980.

139. P. J. E. Forsyth, The Physical Basis of Metal Fatigue, Blackie and Son Limited, London, UK, 1969.

140. R. P. Gangloff, "Crack size effects on the chemical driving force for aqueous corrosion fatigue,"Metallurgical Transactions: A, vol. 16A, no. 5, pp. 953–969, 1985.

141. R. P. Gangloff and R. P. Wei, "Small crack-environment interaction: the hydrogen embrittlement perspective," in Small Fatigue Cracks, R.O. Ritchie and J. Lankford, Eds., pp. 239–264, Metallurgical Society, 1986.

142. M. Goto, "Scatter in small crack propagation and fatigue behavior in carbon steels," Fatigue Fracture of Engineering Materials and Structures, vol. 16, pp. 795–809, 1993.

143. M. Goto, "Statistical investigation of the behavior of small cracks and fatigue life in carbon steels with different ferrite grain sizes," Fatigue Fracture of Engineering Materials and Structures, vol. 17, no. 6, pp. 635–649, 1994.

144. J. C. Healy, L. Grabowski, and C. J. Beevers, "Short-fatigue-crack growth in a nickel-base superalloy at room and elevated temperature," International Journal of Fatigue, vol. 13, no. 2, pp. 133–138, 1991.

145. M. A. Hicks and C. W. Brown, "A comparison of short crack growth behavior in engineering alloys," inFatigue 84, C. J. Beevers, Ed., p. 1337, EMAS, Warley, UK, 1984.

146. S. Hirose and M. E. Fine, "Fatigue crack initiation and microcrack propagation in X7091 type aluminum P/M alloys," Metallurgical Transactions, vol. 14A, no. 6, pp. 1189–1197, 1983.

147. P. D. Hobson, "The formulation of a crack growth equation for short cracks," Fatigue Fracture of Engineering Materials and Structures, vol. 5, no. 4, pp. 323–327, 1982.

148. D. W. Hoeppner, "The effect of grain size on fatigue crack propagation in copper," in Fatigue Crack Propagation, ASTM 415, J. Grosskreutz, Ed., pp. 486–504, American Society for Testing and Materials, 1967.

149. Op. cit. 38.

150. D. W. Hoeppner, "Application of damage tolerance concepts to "short cracks" in safety critical components," in Proceedings of the 12th Industrial College of the Armed Forces Symposium, Industrial Applications of Damage Tolerance (167) Concepts, pp. 2.1/1–2.2/20, ICAF document no. 1336, 1983.

151. T. Hoshide, T. Yamada, and S. Fujimura, "Short crack growth and life prediction in low-cycle fatigue and smooth specimens," Engineering Fracture Mechanics, vol. 21, no. 1, pp. 85–101, 1985.

152. P. Hyspecky and B. Stranadel, "Conversion of short fatigue cracks into a long crack," Fatigue Fracture of Engineering Materials and Structures, vol. 15, no. 9, pp. 845–854, 1992.

153. S. Kawachi, K. Yamada, and T. Kunio, "Some aspects of small crack growth near threshold in dual phase steel," in Short Fatigue

Cracks, K. J. Miller and E.R. de los Rios, Eds., pp. 101–114, Mechanical Engineering Publications, 1992.

154. Y. H. Kim, T. Mura, and M.E. Fine, "Fatigue crack initiation and microcrack growth in 4140 steel,"Metallurgical Transactions: A, vol. 9, no. 11, pp. 1679–1683, 1978.

155. Y. H. Kim and M.E. Fine, "Fatigue crack initiation and strain-controlled fatigue of some high strength low alloy steels," Metallurgical Transactions: A, vol. 13, no. 1, pp. 59–71, 1982.

156. S. Kumai, J. E. Kino, and J. F. Knott, "Short and long fatigue crack growth in a SiC reinforced aluminum alloy," Fatigue Fracture of Engineering Materials and Structures, vol. 13, no. 5, pp. 511–524, 1990.

157. C. Y. Kung and M.E. Fine, "Fatigue crack initiation and microcrack growth in 2024-T4 and 2124-T4 aluminum alloys," Metallurgical Transactions: A, vol. 10, pp. 603–610, 1979.

158. T. Kunio and K. Yamada, "Microstructural aspects of the threshold condition for non-propagating fatigue cracks in martensitic-ferritic structures," in Fatigue Mechanisms, ASTM STP 675, J.T. Fong, Ed., pp. 342–370, 1979.

159. S. I. Kwun and R. A. Fournelle, "Fatigue crack initiation and propagation in a quenched and tempered niobium bearing HSLA steel," Metallurgical Transactions: A, vol. 13, pp. 393–399, 1982.

160. J. Lankford, T. S. Cook, and G. P. Sheldon, "Fatigue microcrack growth in nickel-base superalloy,"International Journal of Fracture, vol. 17, pp. 143–155, 1981.

161. J. Lankford, "The growth of small fatigue cracks in 7075-T6 aluminum," Fatigue Fracture of Engineering Materials and Structures, vol. 5, pp. 233–248, 1982.

162. J. Lankford, "The effect of environment on the growth of small fatigue cracks," Fatigue Fracture of Engineering Materials and Structures, vol. 6, pp. 15–31, 1983.

163. J. Lankford and D. L. Davidson, "The role of metallurgical factors in controlling the growth of small fatigue cracks," in Small Fatigue Cracks, R.O. Ritchie and J. Lankford, Eds., pp. 51–72, The Metallurgical Society, 1986.

164. Y. Mahajan and H. Margolin, "Low cycle fatigue behavior of Ti-6Al-2Sn-4Zr-6Mo: part I. The role of microstructure in low

cycle crack nucleation and early crack growth," Metallurgical Transactions: A, vol. 13, no. 2, pp. 257–267, 1982.

165. R. C. McClung, K. S. Chan, S. J. Hudak, and D. L. Davidson, "Analysis of small crack behavior for airframe applications," in Proceedings of the FAA/NASA International Symposium on Advanced Structural Integrity Methods for Airframe Durability and Damage Tolerance, 1994.

166. A. J. McEvily and S. Minakawa, "Crack closure and the conditions for crack propagation," in Fatigue Crack Growth Threshold Concepts, D.L. Davidson and S. Suresh, Eds., pp. 517–530, TMS-AIME, 1984.

167. Z. Mei and J. W. Morris, "The growth of small fatigue cracks in A286 steel," Metallurgical Transactions: A, vol. 24, no. 3, pp. 689–700, 1993.

168. Z. Mei, C. R. Krenn, and J. W. Morris, "Initiation and growth of small fatigue cracks in a Ni-base superalloy," Metallurgical Transactions: A, vol. 26, no. 8, pp. 2063–2073, 1995.

169. K. J. Miller, "The short crack problem," Fatigue and Fracture of Engineering Materials and Structures, vol. 5, no. 3, pp. 223–232, 1982.

170. K. Minakawa, Y. Matsuo, and A. McEvily, "The influence of a duplex microstructure in steels on fatigue crack growth in the near-threshold region," Metallurgical Transactions: A, vol. 13, no. 3, pp. 439–445, 1982.

171. W. L. Morris, O. Buck, and H. L. Marcus, "Fatigue crack initiation and early propagation in Al 2219-T851," Metallurgical Transactions: A, vol. 7, no. 7, pp. 1161–1165, 1976.

172. W. L. Morris and O. Buck, "Crack closure load measurements for microcracks developed during the fatigue of Al 2219-T851," Metallurgical Transactions: A, vol. 8, no. 4, pp. 597–601, 1977.

173. Y. Murakami and M. Endo, "Quantitative evaluation of fatigue strength of metals containing various small defects or cracks," Engineering Fracture Mechanics, vol. 17, no. 1, pp. 1–15, 1983.

174. D. J. Nicholls and J. W. Martin, "A comparison of small fatigue crack growth, low cycle fatigue and long fatigue crack growth in Al-Li alloys," Fatigue Fracture of Engineering Materials and Structures, vol. 14, no. 2-3, pp. 185–192, 1991.

175. J. C. Newman and P. R. Edwards, "Short crack growth behavior in an aluminum alloy," NATO-AGARD report, An AGARD cooperative test Programme, 1988.

176. M. Okazaki, T. Tabata, and S. Nohmi, "Intrinsic stage I crack growth of directionally solidified Ni-Base superalloys during low-cycle fatigue at elevated temperature," Metallurgical Transactions: A, vol. 21, no. 8, pp. 2201–2208, 1990.

177. J. Z. Pan, E. R. De Los Rios, and K. J. Miller, "Short fatigue crack growth in plain and notched specimens of an 8090 Al-Li alloy," Fatigue Fracture of Engineering Materials and Structures, vol. 16, no. 12, pp. 1365–1379, 1993.

178. S. Pearson, "Initiation of fatigue cracks in commercial aluminum alloys and the subsequent propagation of very short cracks," Engineering Fracture Mechanics, vol. 7, no. 2, pp. 235–247, 1975.

179. J. Petit and A. Zeghloul, "Environmental and nicrostructural influence on fatigue propagation of small surface cracks," in Environmentally Assisted Cracking: Science and Engineering, ASTM STP 1049, W. B. Lisagor, T.W. Crooker, and B. N. Leis, Eds., pp. 334–346, American Society for Testing and Materials, Philadelphia, Pa, USA, 1990.

180. J. Petit, J. Mendez, L. W. L. Berata, and C. Muller, "Influence of environment on the propagation of short fatigue cracks in a titanium alloy," in Short Fatigue Cracks, K. J. Miller and E. R. de los Rios, Eds., pp. 235–250, Mechanical Engineering Publications, 1992.

181. J. Petit and A. Zeghloul, "On the effect of environment on short crack growth behavior and threshold," in The Behavior of Short Fatigue Cracks, K. J. Miller and E. R. de los Rios, Eds., pp. 163–178, Mechanical Engineering Publications, 1986.

182. J. Petit and K. Kosche, "Stage I and Stage II propagation of short and long cracks in Al-Zn-Mg alloys," inShort Fatigue Cracks, K. J. Miller and E. R. de los Rios, Eds., pp. 135–151, Mechanical Engineering Publications, 1992.

183. R. S. Piascik and S. A. Willard, "The growth of small corrosion fatigue cracks in alloy 2024," Fatigue Fracture of Engineering Materials and Structures, vol. 17, pp. 1247–1259, 1994.

184. A. Plumtree and B. P. D. O›Connor, "Influence of microstructure on short fatigue crack growth,"Fatigue Fracture of Engineering Materials and Structures, vol. 14, no. 2-3, pp. 171–184, 1991.

185. J. P. Polak and T. Laskutin, "Nucleation and short crack growth in fatigued polycrystalline copper,"Fatigue Fracture of Engineering Materials and Structures, vol. 13, no. 2, pp. 119–133, 1990.

186. J. M. Potter and B. G. W. Yee, "Use of small crack data to bring about and quantify improvements to aircraft structural integrity," in Behavior of Short Cracks in Airframe Components, p. 18, North Atlantic Treaty Organization, NATO-AGARD, 1983.

187. R .O. Ritchie and J. Lankford, "Small fatigue cracks: a statement of the problem and potential solutions,"Materials Science and Engineering, vol. 84, pp. 11–16, 1986.

188. R. O. Ritchie and S. Suresh, "Mechanics and physics of the growth of small cracks," Berkeley report, University of California, 1995.

189. J. Schijve, "Fatigue crack closure, observations and technical significance," Tech. Rep. number NLR TR-679, National Aerospace Laboratory NLR, Amsterdam, The Netherland, 1986.

190. J. Schijve, "Multiple-site-damage fatigue of riveted joints," in Proceedings of the International Workshop On Structural Integrity of Aging Airplanes, Atlanta, Ga, USA, March, 1992.

191. J. K. Shang, J. L. Tzou, and K. J. Miller, "Role of crack tip shielding in the initiation and growth of long and small fatigue cracks in composite microstructures," Metallurgical Transactions: A, vol. 18, no. 9, pp. 1613–1627, 1987.

192. G. P. Sheldon, T. S. Cook, J. W. Jones, and J. Lankford, "Some observations on small fatigue cracks in a superalloy," Fatigue Fracture of Engineering Materials and Structures, vol. 3, pp. 219–228, 1981.

193. D. Sigler, M. C. Montpetit, and W. L. Haworth, "Metallography of fatigue crack initiation in an overaged high-strength aluminum alloy," Metallurgical Transactions: A, vol. 14, no. 5, pp. 931–938, 1982.

194. R. R. Stephens, L. Grabowski, and D. W. Hoeppner, "The effect of temperature on the behavior of short fatigue cracks in Waspaloy using an in situ SEM fatigue apparatus," International Journal of Fatigue, vol. 15, no. 4, pp. 273–282, 1993.

195. C. M. Suh, J. J. Lee, and Y. G. Kang, "Fatigue microcracks in type 304 stainless steel at elevated temperature," Fatigue Fracture of Engineering Materials and Structures, vol. 13, no. 5, pp. 487–496, 1990.

196. S. Suresh, "Crack deflection: implications for the growth of long and short fatigue cracks," Metallurgical Transactions: A, vol. 14, pp. 2375–2385, 1983.

197. S. Suresh, "Fatigue crack deflection and fracture surface contact: micromechanical model," Metallurgical Transactions: A, vol. 16, no. 2, pp. 249–260, 1985.

198. S. Taira, K. Tanaka, and M. Hoshina, "Grain size effect on crack nucleation and growth in long-life fatigue of low-carbon steel," in Fatigue Mechanisms, ASTM STP 675, J. T. Fong, Ed., pp. 135–173, ASTM, Philadelphia, Pa, USA, 1979.

199. D. Taylor and J. F. Knott, "Fatigue crack propagation behavior of short cracks; the effect of microstructure," Fatigue Fracture of Engineering Materials and Structures, vol. 4, no. 2, pp. 147–155, 1981.

200. K. Tokaji, T. Ogawa, and Y. Harada, "The growth of small fatigue cracks in a low carbon steel; the effect of microstructure and limitations of linear elastic fracture mechanics," Fatigue Fracture of Engineering Materials and Structures, vol. 9, no. 3, pp. 205–217, 1986.

201. K. Tokaji, T. Ogawa, Y. Harada, and Z. Ando, "Limitations of linear elastic fracture mechanics in respect of small fatigue cracks and microstructure," Fatigue Fracture of Engineering Materials and Structures, vol. 9, no. 1, pp. 1–14, 1986.

202. K. Tokaji, T. Ogawa, and Y. Harada, "Evaluation on limitation of linear elastic fracture mechanics for small fatigue crack growth," Fatigue Fracture of Engineering Materials and Structures, vol. 10, no. 4, pp. 281–289, 1987.

203. K. Tokaji, T. Ogawa, S. Osako, and Y. Harada, "The growth behavior of small fatigue cracks; the effect of microstructure and crack closure," Fatigue ‹87, vol. 2, pp. 313–322, 1988.

204. K. Tokaji and T. Ogawa, "The growth of microstructurally small fatigue cracks in a ferritic-pearlitic steel," Fatigue Fracture of Engineering Materials and Structures, vol. 11, pp. 331–342, 1988.

205. K. Tokaji, T. Ogawa, and T. Aoki, "Small fatigue crack growth in a low carbon steel under tension-compression and pulsating-tension loading," Fatigue Fracture of Engineering Materials and Structures, vol. 13, no. 1, pp. 31–39, 1990.

206. K. Tokaji and T. Ogawa, "The effects of stress ratio on the growth behavior of small fatigue cracks in an aluminum alloy 7075-T6 (with special interest in Stage I crack growth)," Fatigue Fracture of Engineering Materials and Structures, vol. 13, no. 4, pp. 411–421, 1990.

207. K. Tokaji, T. Ogawa, Y. Kameyama, and Y. Kato, "Small fatigue crack growth behavior and its statistical properties in a pure titanium," Fatigue ‹90, vol. 2, pp. 1091–1096, 1990.

208. K. Tokaji and T. Ogawa, "The growth behavior of microstructurally small fatigue cracks in metals," inShort Fatigue Cracks, K. J. Miller and E. R. de los Rios, Eds., pp. 85–100, Mechanical Engineering Publications, 1992.

209. K. T. Venkateswara Rao, W. Yu, and R. O. Ritchie, "Fatigue crack propagation in aluminum-lithium alloy 2090: part II. Small crack behavior," Metallurgical Transactions: A, vol. 19, pp. 563–569, 1988.

210. L. Wagner, J. K. Gregory, A. Gysler, and G. Lutjering, "Propagation behavior of short cracks in a Ti-8.6Al alloy," in Small Fatigue Cracks, R.O. Ritchie and J. Lankford, Eds., pp. 117–128, Metallurgical Society, 1986.

211. H. A. Wood and J. L Rudd, "Evaluation of small cracks in airframe structures," personal communication, 1993.

212. D. C. Wu, An investigation into the fatigue crack growth characteristics of a single crystal nickel-base superalloy, Ph.D. dissertation, University of Toronto, 1986.

213. J. R. Yates, W. Zhang, and K. J. Miller, "The initiation and propagation behavior of short fatigue cracks in Waspaloy subjected to bending," Fatigue Fracture of Engineering Materials and Structures, vol. 16, pp. 351–362, 1993.

214. A. Zeghloul and J. Petit, "Environmental sensitivity of small crack growth in 7075 aluminum alloy,"Fatigue Fracture of Engineering Materials and Structures, vol. 8, no. 4, pp. 341–348, 1985.

215. A. K. Zurek, M. R. James, and W. L. Morris, "The effect of grain size on fatigue growth of short cracks,"Metallurgical Transactions: A, vol. 14, no. 8, pp. 1697–1705, 1982.

216. D. W. Hoeppner, et al., "Aircraft structural fatigue," Four volumes of notes, Course sponsored by the US FAA, 1979–1992.

217. M. Creager, T. R. Brussat, D. W. Hoeppner, and T. Swift, "Structural integrity of new and aging metallic aircraft," Short course by UCLA Extension, Department of Engineering, Information Systems and Technical Management, Los Angeles, Calif, USA, 1970-present.

Chapter 2

A Study on Surface Modification of Al7075-T6 Alloy against Fretting Fatigue Phenomenon

E. Mohseni, E. Zalnezhad, Ahmed A. D. Sarhan, and A. R. Bushroa

Department of Mechanical Engineering, Faculty of Engineering, University of Malaya, 50603 Kuala Lumpur, Malaysia

ABSTRACT

Aircraft engines, fuselage, automobile parts, and energy saving strategies in general have promoted the interest and research in the field of lightweight materials, typically on alloys based on aluminum. Aluminum alloy itself does not have suitable wear resistance; therefore, it is necessary to enhance surface properties for practical applications, particularly when aluminum is in contact with other parts. Fretting fatigue phenomenon occurs when two surfaces are in contact with each other and one or both parts are subjected to cyclic load. Fretting drastically decreases the fatigue life of materials. Therefore, investigating the fretting fatigue life of materials is an important subject. Applying surface modification methods is anticipated to be a supreme solution to gradually decreasing fretting damage. In this paper, the authors would like to review methods employed so far to diminish the effect of fretting on the fatigue life of Al7075-T6 alloy. The methods include deep rolling, shot peening, laser shock peening, and thin film hard coatings. The surface coatings techniques are comprising physical vapor deposition (PVD), hard anodizing, ion-beam-enhanced deposition (IBED), and nitriding.

INTRODUCTION

Fretting fatigue phenomenon occurs when two surfaces in contact simultaneously encounter sliding movements and fluctuating loads. Fretting fatigue also occurs when an oscillatory movement with low amplitude between two surfaces is remaining for a large number of cycles [1]. This event can result in two different types of damage: fretting fatigue and fretting wear [2–4]. Bearings, bolted, steel cables, riveted connections and shafts, and steam or gas turbines are common examples of engineering applications facing high fretting fatigue damage risk [5–8]. In contrast to normal fatigue conditions, fretting fatigue may significantly reduce the endurance limit of components. Fretting fatigue may occur in bending, torsion, and even tension forms [9]. Figure 1(a) illustrates the fretting corrosion at an axle-cylinder contact. Figure 1(b) shows fretting wear on the cap screw threads. The fretting fatigue in bolted flanges is presented in Figure 2 [10, 11]. Fretting fatigue is triggered by cracks formed in either surface. It is more serious than fretting wear and fretting corrosion because it can lead to severe component failure.

(a)

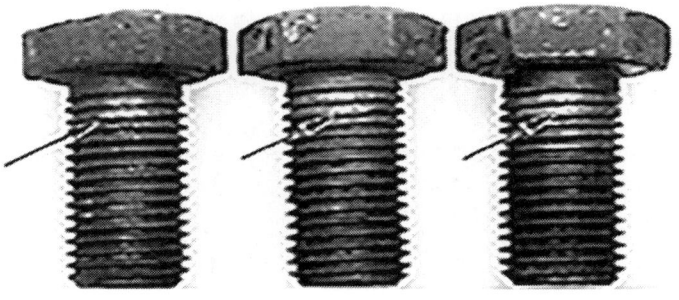

(b)

Figure 1: (a) Fretting corrosion at an axle-cylinder contact and (b) fretting wear on the threads of the caps crews [11].

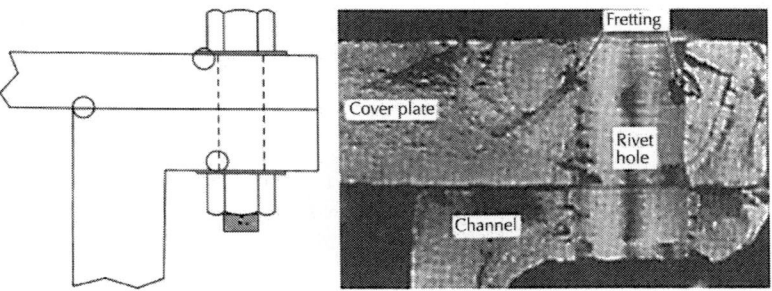

Figure 2: Fretting fatigue in bolted flanges [11].

The fretting fatigue mechanism is schematically arranged and illustrated in Figure 3. Two fretting pads are pushed opposite to the specimen by a load called contact force, P. The component then encounters a cyclic load, Q. Elastic elongation takes place in the sample along the contact zone that causes the fretting fatigue in component. The material's resistivity against fretting fatigue is influenced by contact configuration, slipping amplitude, or surface conditions like lubrication, hardness, friction coefficient, and roughness [12].

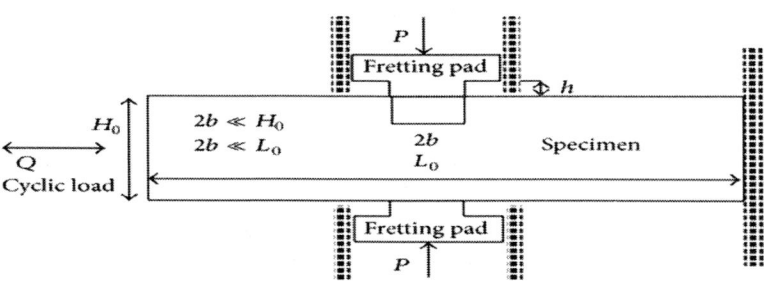

Figure 3: The schematic view of a fretting fatigue arrangement [11].

Generally, three basic means of choosing preventative measures are introduced.

- Change in design: design optimization involves change in contacting materials and component geometry, and it is a basic method of suppressing fretting [13].

- Use of lubricants: using a suitable lubricant (e.g., solid, liquid, or grease) is another possible way of providing practical reduction in fretting damage [14].

- Application of surface engineering: the introduction of surface treatments is likely an ideal solution to diminish fretting damage [15–18].

As mentioned earlier, the fretting fatigue behavior of materials can be significantly enhanced by different surface treatment techniques [19–23] including induction hardening, nitriding, case hardening, shot peening, roll peening, and laser shock peening [24, 25]. The reduction coefficient of fretting fatigue life for the number of engineering alloys is given in Table 1 [26]. The material's durability to fretting fatigue is significantly increased by enhanced surface conditions such as roughness, hardness, and lowered friction coefficient, something achievable by surface treatment techniques like surface-coatings, deep rolling, and shot peening that may postpone crack initiation.

Table 1: The reduction coefficient of fretting fatigue life for some engineering alloys [26]

Fatigue specimen	Contact pad	Contact pressure (MPa)	Fatigue mean stress (MPa)	Fatigue strength unfretted (MPa)	Fatigue strength fretted (MPa)	Strength reduction factor	Relative slip range (m)
Steels							
3.5NiCrMoV	1CrMo	30	0	±300	±140	2.1	15
3.5NiCrMoV	1CrMo	30	300	±215	±60	3.6	6.2
3.5NiCrMoV	1CrMo	300	0	±300	±130	2.3	12
3.5NiCrMoV	1CrMo	300	300	±215	±60	3.6	4
3.5NiCrMoV	2014 A	30	0	±300	±140	2.1	1.5
3.5NiCrMoV	2014 A	30	300	±215	±75	2.9	6.8
18M114Cr	3.5NiCrMoV	20.7	0	±250	±100	2.5	10
18M114Cr	3.5NiCrMoV	20.7	300	±125	±50	2.5	4.5
18M114Cr	3.5NiCrMoV	20.7	0	±250	±165	1.5	17.4
18M114Cr	3.5NiClMoV	20.7	300	±185	±70	2.6	6.7
Aluminum							
2014A Al	3.5NiCrMoV	30.8	75	±140	±15	9.3	2
2014A Al	3.5NiCrMoV	30.8	125	±135	±12.5	10.8	1.5
2014A Al	3.5NiCrMoV	30.8	125	±135	±50	2.7	9.6
Peened Aluminum							
2014A	BS S98	103.5	0	±148	±72	2.05	2
2014A	BS S98	103.5	0	±148	±47	3.14	4.2
2014A	BS S98	103.5	0	±148	±36	4.11	8.35
2014A	BS S98	103.5	0	±148	±36	4.11	17.4
Titanium							
Ti6Al4V	Ti6Al4V	20	61.25	±260	±125	2.1	30

Improving the mechanical properties of surface by leaving the treated region in compressive residual stress states has the major benefit [27–32].

Automobile parts, fuselage, aircraft engines, and energy saving strategies generally promote the research interest in the area of lightweight materials by mainly aluminum alloys. The pure aluminum is not suggested to be used as structural parts since it does not result in satisfactory mechanical strength. Hence, in practical applications, it is essential to improve the surface properties, especially prior to aluminum coming in contact with other parts [33, 34].

Aerospace structural fastened joints utilize Al 7075-T6 and fretting fatigue damage can result in catastrophic failures under fluctuating loading [35, 36]. Aluminum alloy 7075-T6 has high ratio of strength to weight and low specific weight besides high thermal and electrical conductivity. The formation of thin hard coatings on material surfaces is one of the best ways to enhance the material wear resistance. Hard coatings also appear to be encouraging by means of the feasibility of approaching high strength, hardness, and concurrently high protective and decorative surface properties [37].

This study focuses on the variety of surface treatments available to mitigate the fretting fatigue of aluminum 7075-T6. Deep rolling, shot peening, TiN coating, CrN coating, and nitriding surface treatments are discussed.

SURFACE TREATMENT USING MECHANICAL TECHNIQUES

Deep Rolling

Deep rolling (DR) includes a ball or roller type tool to induce surface-compressive residual stress. Such stress enhances the fatigue resistance of engineering components and materials. A deep rolling fixture is illustrated in Figure 4. This technique is distinct from roller burnishing whose main purpose is to achieve an exceptionally well-polished surface. The ball or roller produces a longitudinal groove as it comes in contact with the component's surface. A plastic region along with an elastic zone couples with this groove. Upon separation from the roller, a large, compressive residual stress is produced on the surface as a result of recovering the elastic zone. Surface rolling residual stresses are generally consequent to the interface between the plastic and elastic regions that result from the pressure between the roller and the contact surface. Specific parameters can considerably affect the process of deep rolling followed by near-surface residual stress; force of rolling is recognized to be the main one. Very low level of rolling forces has no serious effect on the behavior of fatigue, but considerably high force may make it even worse, for example, by initiation of microcracks. As such, it appears that only optimized rolling forces are able to promote fatigue strength [38].

Figure 4: The fixture used for DR the specimens [38].

The aero and automobile industries (in particular crankshafts) commonly utilize the DR technique to develop fatigue resistance. Compared with other methods, DR has two distinct advantages: (i) lower surface roughness and (ii) greater compressive residual stress depth.

Shot Peening

Shot peening (SP), for decades, was known as a surface treatment with questionable benefits about cyclic loading [39]. The inconsistent results were partially due to ignorance with respect to the shot peening process and partly result from a lack of proper background that would allow the characterization of the role that surface modifications produced by shot peening play in fatigue damage. Nowadays, the control parameters of shot peening performance, such as intensity, media, and coverage, are well understood, and new designation of controlled shot peening (CSP) has emerged. The CSP or SP is a cold work process accomplished by bombarding the workpiece surface with small-diameter ferrous and nonferrous spherical shots. A schematic of the shot peening process to induce compressive residual stress is shown in Figure 5.

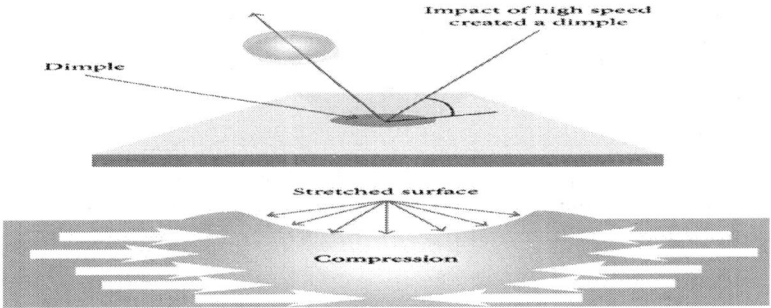

Figure 5: Schematic of shot peening bombardment on a surface with small high quality spherical media [5].

SP is widely applicable for enhancing the industrial component's fatigue behavior, particularly in the car industry [40–43]. SP even acts as a forming process in the production of large, thin aero industrial components including wide panels. SP is known to harden a material's surface and consequently increase its strength. The surface modifications yielded by SP are (a) surface roughening; (b) an increased, near-surface, dislocation density (strain hardening); and (c) the development of a characteristic residual stress profile [44–46].

Laser Shock Peening

Laser shock peening (LSP) is a promising surface treatment technique which has demonstrated effectiveness in enhancing the fatigue properties of a number of metals and alloys. The process was originally developed at the Battelle Columbus Laboratory in the 1970s [47–51]. The LSP procedure is illustrated schematically in Figure 6. Since then, considerable attention has been directed to potential LSP applications in the aerospace and automotive industries. The beneficial effects of LSP on the static, cyclic, fretting fatigue and stress corrosion performance of aeronautical and automotive aluminum alloys, steels, and nickel-based alloys have been confirmed [52–59]. LSP has also been successfully engaged in upgrading the resistance of aircraft gas turbine engine blades to foreign object damage. Since laser beams can be easily directed to fatigue-critical areas, LSP technology is expected to be widely applicable to improving the fatigue properties of metals and alloys, particularly those that respond positively to shot peening.

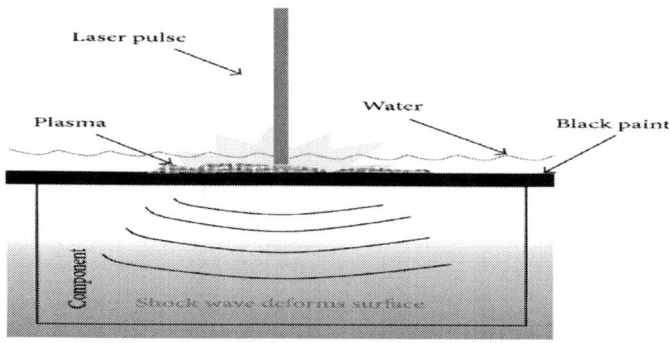

Figure 6: Schematic of the laser peening process.

MECHANICAL TECHNIQUES FOR IM-PROVING THE FRETTING FATIGUE LIFE OF AL 7075-T6

Deep Rolling

Majzoobi et al. [38] looked into how deep rolling (DR) affects the fretting fatigue life of Al 7075-T6. In order to assess the effectiveness, they compared shot peening (SP) with deep rolling (DR) on the fretting fatigue behavior of Al 7075-T6. In their study, the following experiments were examined: (i) fretting fatigue tests on roll-peened specimens with high force of rolling, (ii) fretting fatigue tests on intact specimens, (iii) fretting fatigue tests on shot-peened specimens, (iv) normal fatigue tests on virgin specimens, and (v) fretting fatigue tests on roll-peened specimens with low force of rolling.

The experimental results are graphically illustrated in Figure 7, where it is evident that the fretting fatigue tests on intact samples (with no surface treatment) produced the lowest fatigue life. However, the situation differed in the case of the surface-treated samples. The bar chart in Figure 8 depicts the influence of different surface treatment techniques on the fretting fatigue behavior of the material. Also, apparently, the results (Figure 7) indicate that fretting fatigue decreases normal fatigue life at a stress of 130 MPa by roughly 67%. The fatigue life reduction rate, though, diminishes with rising stress [38].

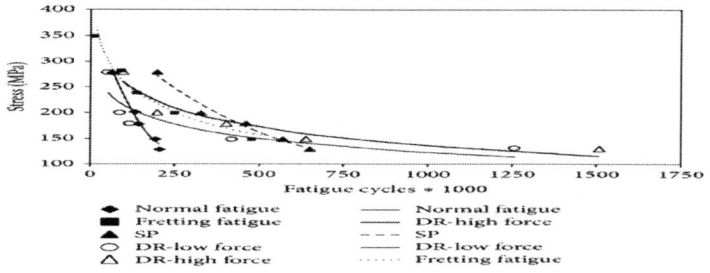

Figure 7: S-N curves for various fatigue testing conditions [38].

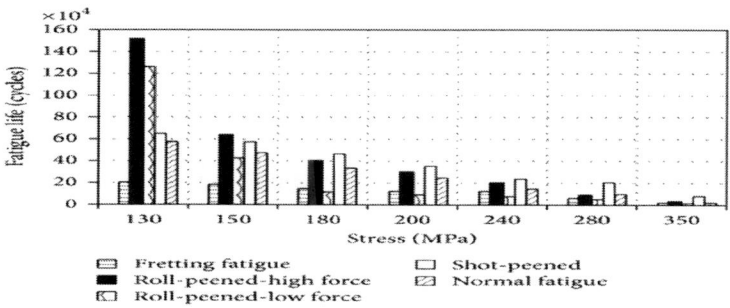

Figure 8: Bar chart for various fatigues testing conditions [38].

The profiles of residual stress formed by DR in the x, y, and z directions are shown in Figure 9 for the roller's linear motion xx, the roller's transverse displacement yy, and load zz (Figure 10). Obviously, the stress in the y-direction is more significant than the other two components, and it runs perpendicularly to the rolling direction. It should be mentioned that the oscillatory motion of contacting surfaces in the fretting fatigue tests, for example, between the specimen and pad, is in the x-direction [38].

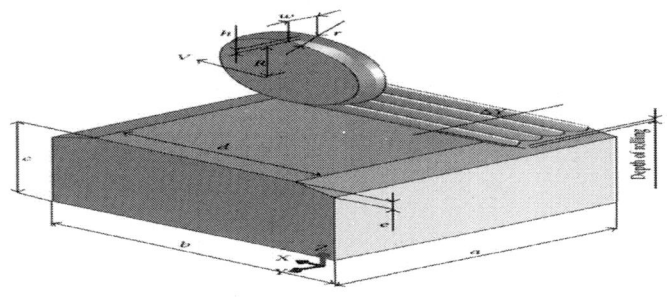

Figure 9: Residual stress profiles induced by DR in three directions [38].

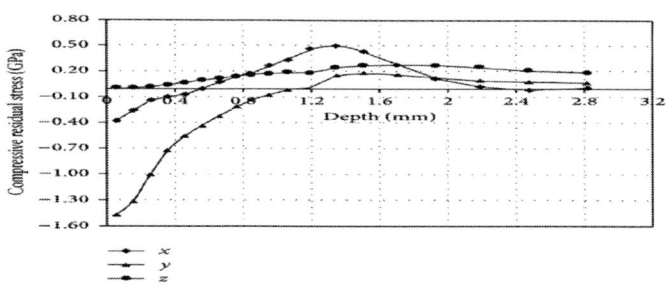

Figure 10: The model of deep rolling (not scaled) used in the simulations [38].

Results provide evidence that for low cycle fatigue (LCF) SP remains exceptionally high, up to approximately 300,000 cycles. For more than 300,000 cycles, the influence of DR on resistance to fretting fatigue is superior to SP where fatigue cycles increase by 700% (at 130 MPa stress) at high rolling force deep rolling. The rate of increase in fatigue life slows down as the applied stress diminishes so that for the stress of 150 MPa the fretting fatigue life increases about 400%, while for the stress of 280 MPa the fatigue life increase is only about 50% [38].

Shot Peening

Much like DR, SP produces a layer of compressive residual stress near the workpiece surface. The distribution and depth of residual stress are potentially affected by a variety of parameters such as material, shot diameter, target material, duration, intensity of exposure, and shot velocity. SP is widely utilized for enhancing the fatigue life of industrial components, more so in the car industry [40–43].

Majzoobi et al. [38] studied the influence of shot peening and deep rolling on the fretting fatigue life of Al 7075-T6. Based on their results, which are illustrated in Figures 7 and 8, the SP technique maintains a superior performance position to DR at low cyclic fatigue (LCF). The shot peening technique improves fretting fatigue life by a significant 300%. Majzoobi and Ahmadkhani in a similar endeavor investigated the influence of regular reshot peening on the resistance of Al7075-T6 against fretting fatigue [60]. They found that it is possible to considerably enhance the material's behavior against fretting fatigue with reshot peening. After 5 reshot peening processes, the fretting fatigue life of Al7075-T6 improved by 600%. Figure 11 presents the variation of fretting fatigue life versus maximum stress for each stage of reshot peening. It is clear that the gap between the curves becomes wider at less stress, which is supported by the bar chart in Figure 12 where fretting fatigue life improvement at lower stresses is more considerable than at higher stresses [60].

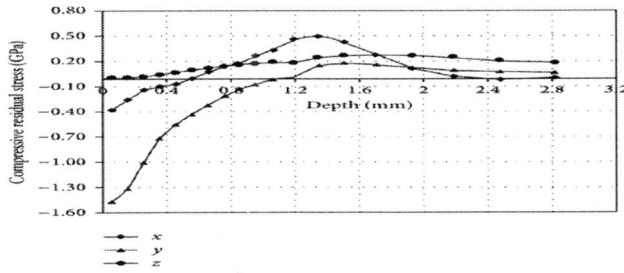

Figure 11: Variation of fatigue life versus maximum stress for multiple reshot peening [60].

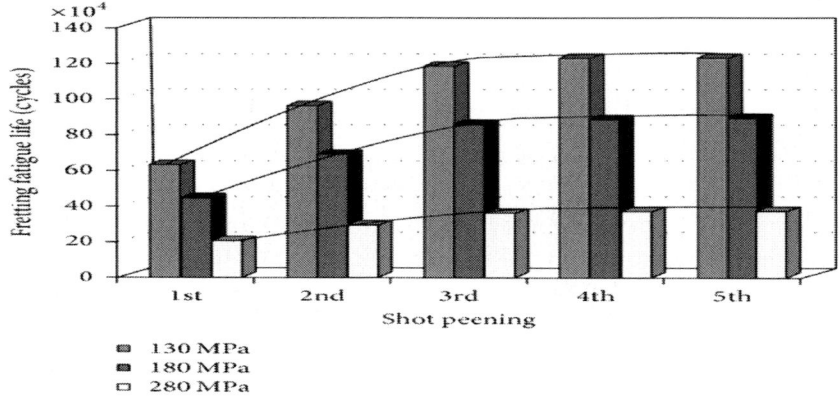

Figure 12: Variation of fatigue life versus the number of reshot peenings for different maximum stresses [60].

Reshot peening is more efficient during initial processes. For example, an almost 100% increase in fatigue life was observed for a first round of reshot peening. This increase is logarithmically decreased for subsequent reshot peenings. It is reported that the 5th reshot peening produced a negligible improvement of less than 2%.

The improvement in fatigue life as a result of reshot peening can be credited to (i) the closing of small cracks that had already initiated and grown on the component's contacting surface, (ii) removal of debris created by wear and the fretting phenomenon from the shot peening process, (iii) the recreation of a layer of compressive residual stress through the reshot peeing process, and (iv) a combination of reasons (i) to (iii) [60].

From a comparative study between DR and SP, it is concluded that (i) at low cyclic fatigue the SP technique improves the material's fretting fatigue behavior more efficiently than DR; (ii) a 300% improvement has been attained for the SP-tested specimen; however, (iii) the influence of DR on fretting fatigue resistance proved more effective for high cycle fatigue; for example, there was an increase of up to 700% for DR at higher forces of rolling; (iv) with decreasing the force of rolling, the trend of increase in fatigue life slowed down; (v) the surface modification technique has direct impact on the friction coefficient such that the highest and lowest friction coefficients, 0.70 and 0.45, were achieved for the low force deep-rolled and intact sample, respectively.

Laser Shock Peening

The resistance of alloys and metals to fretting fatigue and fatigue is normally increased by laser shock peening (LSP) and laser peening [61]. The boost in resistance is accomplished by a high energy pulsed laser that creates residual compressive stresses and strain hardening into the surface of the part being laser peened. Compared with shot peening, the compressive residual stresses produced by laser shock peening penetrate deeper into the surface than those obtained by shot peening. Thus, significantly greater fatigue resistance promotion is noted after laser shock peening.

Laser shock peening is the most effective against fretting fatigue [62]. Dog-bone specimens and pads of aluminum 7075-T6 alloy were laser treated around a simulated fastener hole located in each piece (Figure 13); the pieces were then fixed firmly together through the holes with a manufactured fastener. The fatigue test at R=0.1 was carried out on this combination. The stress differential between the smaller cross-section of the dog-bone and the larger cross-section of the pad created an elongation differential between each piece during a cycle, leading to fretting around the fastener hole. The results are provided in Figure 14. The tests were initially conducted at 96 MPa. Upon achieving a lengthy life, the stress was increased by 10% until failure occurred after a few hundred thousand cycles. LSP expanded fretting fatigue life even at 113 MPa.

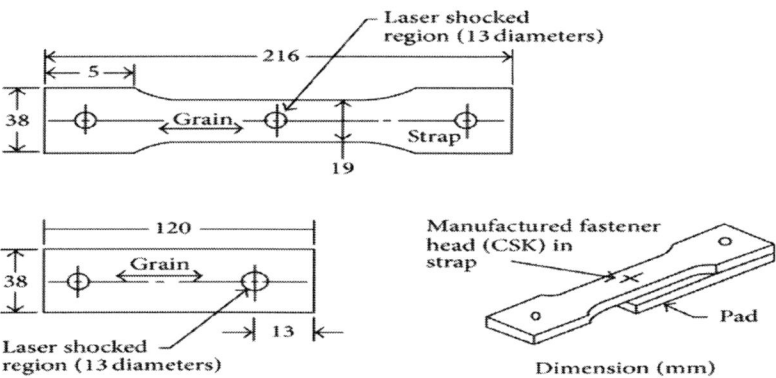

Figure 13: Ferreting fatigue specimen configuration [62].

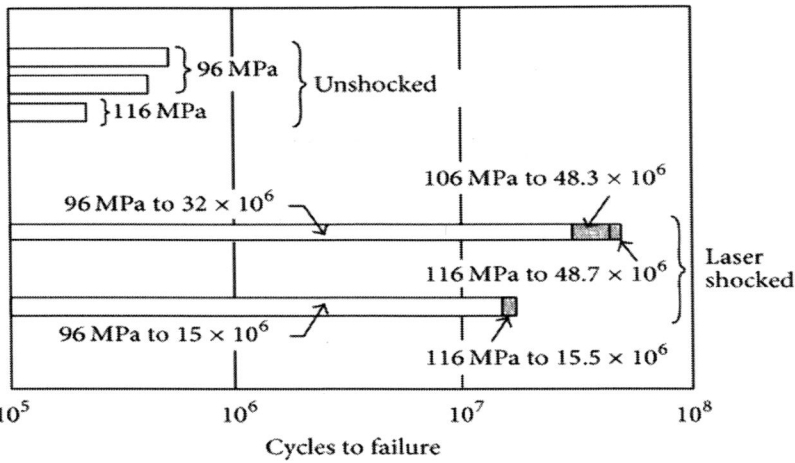

Figure 14: Increased resistance to ferreting fatigue around fastener holes after laser shock peening aluminum 7075-T6 [62].

Earlier, it has been discussed that SP is a successful technique, whereby the component surface is deformed plastically using multiple overlapping impacts of ceramic or metal shot [63, 64]. However, the depth of the compressive residual stress is practically only about 250 μm, maintaining low fretting fatigue resistance. Accordingly, King et al. [52] demonstrated a comparison between SP and LSP. Compared to SP, LSP is able to produce compressive stresses to greater depths, ~1.5 mm, while at the same time keeping lower work hardening levels [65–67].

King et al. [52] claimed that the fretting fatigue loading of dovetail biaxial rig samples with contact surfaces treated with combined SP and LSP results in considerable stress relaxation that penetrates 0.5 mm deep. The penetration of compressive residual elastic strain (~1.5 mm) is unaffected. Their study reports that fretting results in plastic relaxation of the misfit parallel to the fretting direction which extends roughly 0.4 mm below the fretted and laser peened surface, despite the less significant misfit relaxation that is usual to the fretting direction. Basically, the extended depth of compressive residual stress produced by laser shock peening as opposed to shot peening (1.5 mm instead of 0.3 mm) is a significant consideration in enhancing the tolerance of components to fretting damage.

COATING SURFACE TREATMENTS

Physical Vapor Deposition (PVD)

PVD coatings are popular in many cutting applications mostly due to their high hardness [68, 69]. Other applications are known to take advantage of these coatings for upgrading the fretting fatigue and wear resistance of contacting components [70–72]. With the advent of modern technologies like vacuum processing, high power laser, and progress in materials such as composites and ceramics, surface modification methods based on current technologies have found additional demand with respect to traditional surface modifications ranging from painting and glazing to electroplating and gas carburizing over the past decade [22]. Vacuum coating procedures carry the potential to apply higher-hardness coatings than any metal. PVD is one of the vacuum coating techniques known to condensate from vapor phase to solid phase, having the film material deposited atom by atom on the substrate. PVD metal coatings like silver and gold were conventional means of avoiding fretting failure [73]. Studies have reported that under fretting amplitudes of about 50 ± 100 mm metal coatings like zirconium and chromium make durable films [74]. This technology permits coating deposition at temperatures as low as 200°C (390°F). Maintaining a low temperature facilitates the coating of materials with neither loss of hardness, distortion nor reduction in corrosion resistance, while PVD coatings experience no performance deterioration in comparison with materials deposited at higher temperatures. Improved surface hardness and higher service temperatures are obtainable with PVD [75–77]. Thermal evaporation, ion plating, and sputtering are the three main techniques of applying PVD coatings [78]. More recently, PVD coating has become probably one of the most widely used and successful sorts of coating meant to improve fretting fatigue life [79].

Hard Anodizing

Hard anodize coating is a very effective surface treatment employed with the aim of reducing the destructive influence of fretting fatigue. Hard anodizing is encouraging to obtain high strength and hardness, while simultaneously presenting fine surface decorative and protective properties. This coating resulting by an electrochemical process performs stable oxide layers on the surface of the metal. Anodic coating can be deposited to surfaces of aluminum through different electrolytes with AC, DC, or a combination of both, to enhance the metal hardness. Hard-anodized coating is not accepted by all types of aluminum alloys and thus is not applicable in such cases. Aluminum alloys with high silicon or copper content are prone to not being very hard and porous. Table 2 lists some of the alloys that are difficult to be coated and should be practically avoided [80–83]. Pure aluminum coating via magnetron sputtered Al7075-T6 alloy may facilitate substrate hard anodizing.

Table 2: The list of the aluminum alloys which should be avoided to hard anodizing [83].

Difficult Al alloys for hard anodizing
2011
2017
2024
7075
Cast and wrought alloys with
Cu > 4% or Si > 7%

Optimizing the effect of each parameter during the hard anodizing process is a critical requirement to improve surface hardness [84, 85]. Anodizing equipment (Figure 15) comprises electrolytic solution, a power supply, cathode (stainless steel), and anode (substrate material). A thin oxide layer is formed on account of the reaction between the oxygen and the substrate; the layer is abrasion-resistant and durable, while hydrogen is produced at the cathode. The anodizing generates uniform oxidation layer (coating) that is much denser and harder than natural oxidation. Moreover, thanks to the hard anodizing coating, the substrate melting point is augmented from about 650°C to approximately 2000°C, which is enough to sustain the mechanical properties at elevated temperatures.

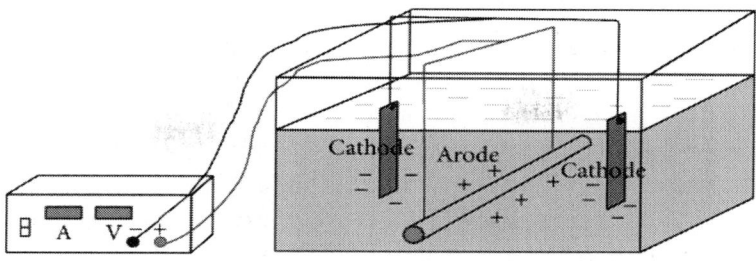

Figure 15: The schematic of hard anodizing process [83].

Ion Beam Enhanced Deposition (IBED)

Ion beam enhanced deposition (IBED) is a process in which PVD coated films are bombarded by an independently produced flux of ions at the same time [86, 87]. Atomic displacements at the bulk and the surface and the amplified migration of atoms within the surface are two results of the massive energy applied to the deposited atoms [88]. The ensuing atomic motions are responsible for improving the film properties as opposed to similar films generated by PVD in the absence of ion bombardment. The advantages of ion-beam-assisted deposition are low deposition temperature, control of stress level, high density, control of microstructure (nanocrystaline, metastable crystalline, or amorphous), precise modulation of composition with depth and high versatility for metals, ceramics, semiconductors, and dielectrics. The limitations of ion-beam-assisted deposition are relatively high cost and line-of-sight [89]. When ions are reactive, compounds such as TiN, Si_3N_4, and BN can be synthesized at fairly low temperatures [90–92]. Thin films of varying stoichiometry or functionally graded thin films can also be produced by adjusting the ratio of reactive ions to atoms reaching the substrate surface [93].

Nitriding

Nitriding is a heat-treating process that diffuses nitrogen into a metal surface to create a casehardened surface. It is commonly used on steel, aluminum, titanium, and molybdenum. The major types comprise ion, plasma, laser, and gas nitriding.

Plasma nitriding is a thermochemical treatment method with several advantages such as control of nitrided layer depth and phase formation [94, 95]. This kind of treatment requires special equipment and high ionizing energy. The two major plasma processes developed for titanium nitride synthesis are ion nitriding and PVD [96]. One disadvantage of plasma nitriding is that it reduces the fretting fatigue strength of aluminum alloys, a problem that can nevertheless be overcome by lowering the processing temperature [97]. Ion-beam nitriding, which includes the high-energy end spectrum, is an alternative method of hardening aluminum and titanium alloy surfaces. The treated surface is encountered to the ion beam using N_2 and Ar [98, 99].

Laser nitriding melts the surface, up to 1.5 am deep, using a focused laser beam in a nitrogen gas environment to produce a hard layer of titanium nitride. Surface cracking is the main drawback with laser nitriding of titanium alloys [100]. The main disadvantages are special equipment requirement and dependency on the material's geometry.

Gas nitriding is deemed a promising method for engineering applications, as it forms harder layers on material surfaces effortlessly. Gas nitriding has the key benefit of being independent from the sample's geometry and it does not involve special equipment. A great disadvantage entails the high temperatures necessary, around 650–1000°C, as well as lengthy nitriding times of 1–100 h, as reported in the literature. It is also common knowledge that gas nitriding reduces the fretting fatigue limit of aluminum alloys [101].

Plasma nitriding or ion nitriding is more preferred than gas nitriding because of advantages such as the ability to select either an   or a monophase layer or even to remove the lower treatment, white layer, temperatures, and better control of thickness of the case [102]. The heat treatment makes it easier to control the dimensions and in some cases eliminate machining altogether [102]. Fatigue strength is notably improved by nitriding. The formation of precipitates in the diffusion layer tends to increase the hardness and create compressive residual stresses. These beneficial stresses lower the magnitude of the applied tensile stresses and hence increase the fretting fatigue life of the component.

APPLYING DIFFERENT TYPES OF COATINGS TO IMPROVE THE FRETTING FATIGUE LIFE OF AL 7075-T6

PVD Coating

Puchi-Cabrera et al. [103] explored the effect of TiN coating on Al 7075-T6 substrate by magnetron sputtering PVD coating technique. The TiN-PVD coating layer contributed to considerably enhancing the fretting fatigue life of the Al 7075-T6 substrate by 400% up to 2119% depending on the maximum alternating stress applied to the material. The improvement in fatigue life was a result mainly of the compressive residual stresses within the coating of roughly 7.08 GPa and the excellent adhesion between the coating and the substrate. It was also reported that TiN-PVD coatings maintain fine adhesion to the substrate under fretting fatigue loads [103].

Zalnezhad et al. studied the fretting fatigue life of Al7075-T6 by application of TiN coating using PVD magnetron sputtering technique with the highest adhesion strength and surface hardness as well as minimum surface roughness achieved by improving the parameters of TiN coating [104]. Reportedly, fretting decreases the fatigue life for uncoated Al7075-T6 by 30% in a low cycle region and 57% in a high cycle region. The fretting fatigue lives of TiN-coated samples with high surface hardness and high adhesion increased at high cyclic fatigue by 39% and 77% and low cyclic fatigue by 61% and 16%, respectively, in comparison with the uncoated samples.

From the study of Figures 16 and 17, it can be concluded that the fretting fatigue life of the TiN-coated specimens with high adhesion enhanced more than the coated specimen with higher surface hardness at low cyclic fatigue, while at high cyclic fatigue the results are reversed. This result is due to the brittleness of the TiN hard coating, where at high cyclic fatigue the cracks are propagated in the coating and then developed inside the specimen.

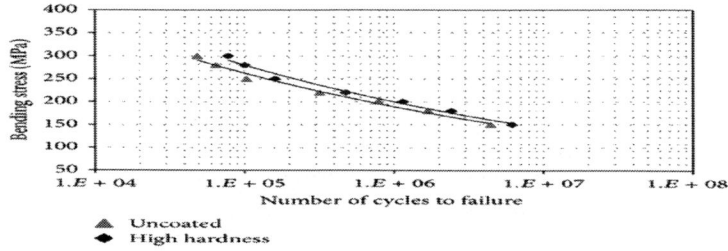

Figure 16: S/N curve of fretting fatigue for uncoated and TiN coated specimens with highest surface hardness [104].

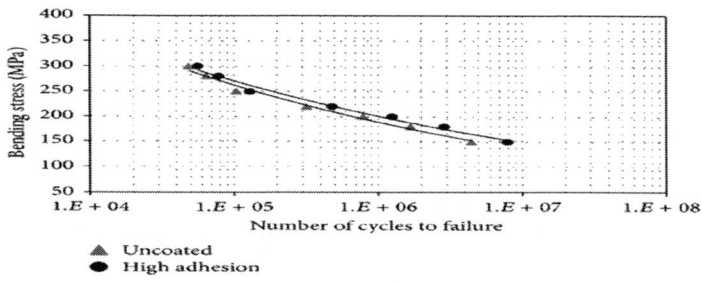

Figure 17: S/N curve of fretting fatigue for uncoated and TiN coated specimens with highest surface adhesion [104].

In another undertaking [105], ZrN-PVD coatings with 3 μm thickness seemed to experience drastic decline in both the tensile and fatigue properties of an Al 7075-T6 substrate. Decrease values of 28% and 43% were noted in the ultimate strength and yield strength of aluminum substrate, respectively, besides 73% to 82% reduction in fretting fatigue life. The adhesion between the coating layer and substrate appeared good under tensile loading, whereas major delamination occurred under high fatigue loads [105].

Hard Anodizing

Sarhan et al. [83] studied how the fretting fatigue life of hard-anodized aluminum alloy (AL7075-T6) can be influenced by surface hardness by a series of rotary bending fatigue tests. As previously stated, pure aluminum is incapable of accepting hard anodizing. For this reason, pure aluminum was coated on the AL7075-T6 surface. Taguchi optimization method was employed to study the influence of hard-anodized coating parameters including temperature, voltage, time, and solution concentration on surface hardness. Figures18 (a) and 18(b) demonstrate the fretting fatigue life of hard-anodized and uncoated samples with 393 and 360 HV surface hardness, respectively. Figure 18(a) indicates that at low bending stress the fretting fatigue life of hard-anodized specimens with hardness of 360 HV increased in comparison with uncoated specimens. Nonetheless, from 250 to 300 MPa, the results were reversed at high bending stress. Figure 18(b) illustrates the fretting fatigue life of uncoated and hard-anodized specimens with 393 HV as the highest surface hardness obtained. The fretting fatigue lives of hard-anodized samples increased only at low bending stresses from 150 to 200 MPa, but at a hardness of 393 HV fretting fatigue life decreased when bending stress increased from 200 to 300 MPa. Figures 18(a) and 18(b) also indicate that the fretting fatigue lives of hard-anodized samples with surface hardness of 360 HV increased even at 220 MPa bending stress, while at this bending stress the fretting fatigue lives of hard-anodized specimen with 393 HV hardness are decreased compared to uncoated samples. This event is due to the microcracks generation and brittleness (ceramic) of the hard-anodized specimen with higher surface hardness (higher hardness causes higher brittleness) [106–108].

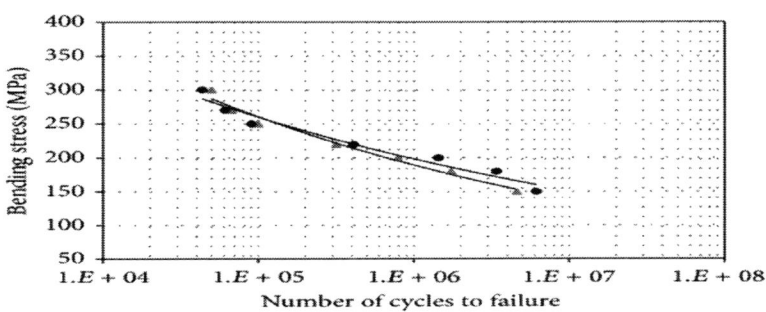

(a)

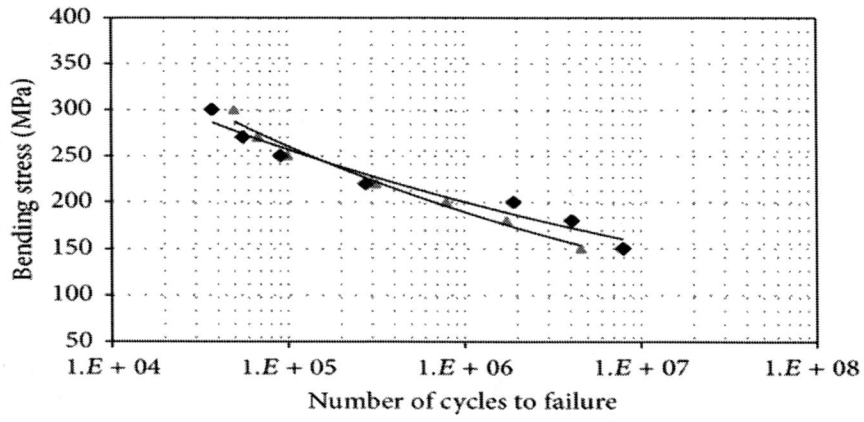

(b)

Figure 18: Comparison of S/N curve of fretting fatigue for uncoated and hard-anodized specimen. (a) S/N curve of fretting fatigue for uncoated and hard-anodized specimens (360 HV) and (b) S/N curve of fretting fatigue for uncoated and optimum hard-anodized parameters specimens (393 HV) [83].

It is supposed that cracks of fretting fatigue appear in regions with frictional shear stress that is locally focused on the contacting surface. Therefore, the diminishing fatigue life resulting by fretting damage is assumed to be due to shorter crack initiation life owing to concentration of local stress caused by fretting, along with accelerated propagation of initial cracks by fretting [109]. One of the principle mechanisms in the acceleration of initial cracks by fretting is considered to be the wedge effect whereby wear debris moves into the small, initial fretting fatigue crack [110]. If the crack is already completely filled with wear debris, the effect tends to be reduced because the wear debris is unable to enter the crack any more [83].

Considerable improvement is achievable through a hard anodizing surface treatment layer, which has the potential to be applied to Al 7075-T6 and has the benefit of pure aluminum as an initial layer coated by PVD magnetron sputtering. However, this improvement is conditional to stress level.

IBED Technique

The fretting resistance and mechanisms of IBED CrN films have been assessed by Fu et al. [111]. The results were weighed against those of CrN films coated onto Al 7075-T6 via PVD. The researchers claimed that the hard IBED CrN film with sufficient substrate adhesion is capable of decreasing ploughing and subsurface distortion, consequently increasing the fretting fatigue resistance considerably, especially during operation. Also, the CrN phase can enhance the antioxidative and anticorrosive resistance of Ti alloy substrate during long-term fretting [79, 111].

One of the rationales behind IBED CrN thin film coating technique for improving fretting fatigue resistance making it superior compared to CrN film by PVD is likely attributed to the generation of a comparatively dense and fine structure due to ion beam bombardment throughout the IBED process [86, 87]. The high adhesion strength of IBED films to the substrate is another potential explanation for the higher fretting fatigue resistance [88, 90, 91]. Under high normal load and very small slip amplitude, long cracks are generated on the surface of the fretting scar on IBED CrN films, representing fretting fatigue failure [110].

Nitriding

Nitriding treatment on Al 7075-T6 alloy has been studied by Majzoobi and Jaleh [22]. They compared fretting fatigue tests (a) without any surface modification, (b) nitriding in the absence of substrate temperature control, (c) nitriding with substrate temperature control, (d) titanium coated by magnetron sputtering PVD technique, (e) titanium coated by ion coating technique, (f) nitriding followed by titanium coating using magnetron sputtering PVD technique, and (g) nitriding followed by titanium coating done by ion coating technique. The relation of fretting fatigue versus maximum stress for the untreated aluminum alloy is illustrated in Figure 19. It is obvious that the drop in fretting fatigue life develops as long as the applied stress increases. For instance, at a stress of about 150 MPa, the decrease is approximately 60%.

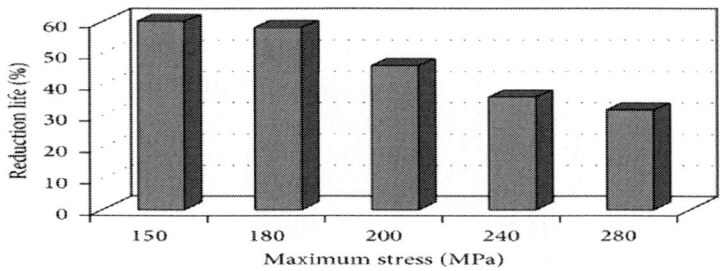

Figure 19: Reduction percentage of fretting fatigue life versus stress [22].

The fatigue test result regarding the surface modified specimens, illustrated in Figure 20, reports that fretting fatigue life for intact samples with no surface modification is much higher than nitrided samples. A couple of test series were conducted on nitrided samples in their study. In the first series of the tests, the temperature was not controlled but allowed to reach about 350°C during nitriding. The outcome of this series is designate by HT (high temperature) in Figure 20. In the second test series, the temperature increase along surface modification was monitored by cooling the specimens. The result is indicated by LT (low temperature) as shown in Figure 20. It can be concluded that the fatigue life drastically shortened for the first series of nitrided specimens. This potentially results from the nature of the nitriding process itself, which dramatically alters the material properties, particularly its ultimate strength and yield stress. Therefore, it can be deduced that nitriding alone does not enhance the material's fretting fatigue behavior.

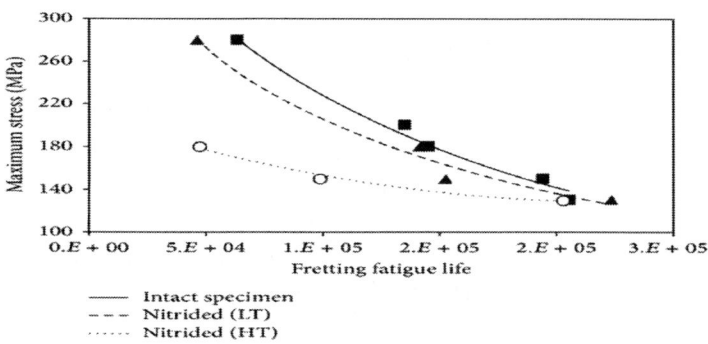

Figure 20: S-N curves for nitrided and normal specimens [22].

Majzoobi and Jaleh [22] have also studied the outcome of titanium-coated specimens and determined that none of the nitriding and titanium coating satisfactorily improved the fretting fatigue response of AL7075-T6. For this reason, further investigation was done on the influence of a duplex surface treatment technique including titanium coating plus nitriding on the Al7075-T6 fretting fatigue behavior. Two test series were assessed. The first one was conducted with the specimens initially nitrided under controlled temperature conditions followed by coating using magnetron sputtering PVD technique. The results are illustrated in Figure 21. The figure clearly demonstrates that the duplex surface treatment technique significantly enhanced the material's fretting fatigue life in testing condition. It was also observed that the influence of the duplex treatment is much better at lower stress. As an example, the fretting fatigue life improved by nearly 100% at a stress of 130 MPa, whereas at a stress of 280 MPa it improved only by 13%.

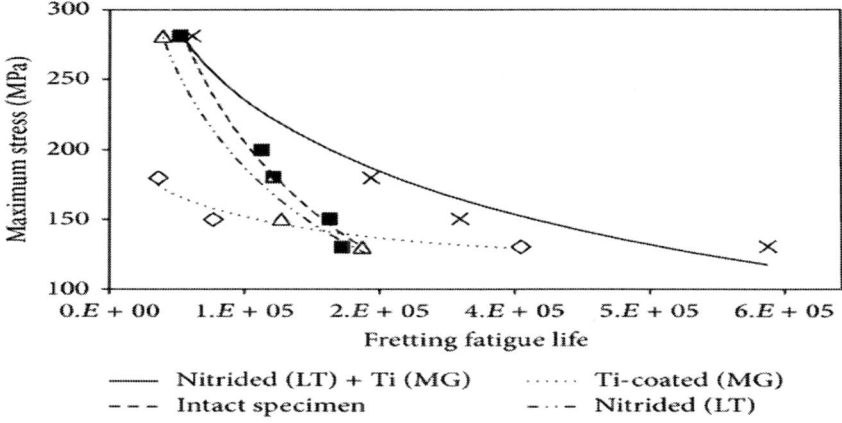

Figure 21: S-N curves for the duplex surface treatment, (Ni (LT) + Ti (MG)) [22].

In the second test, nitriding was followed by coating via hollow cathode deposition technique. As reported in Figure 22, a comparison of the S-N curves in the nitriding with titanium ion coating duplex treatment with the single nitriding treatment suggests that the former does not prompt any change in the performance of material with respect to resistance of fretting fatigue.

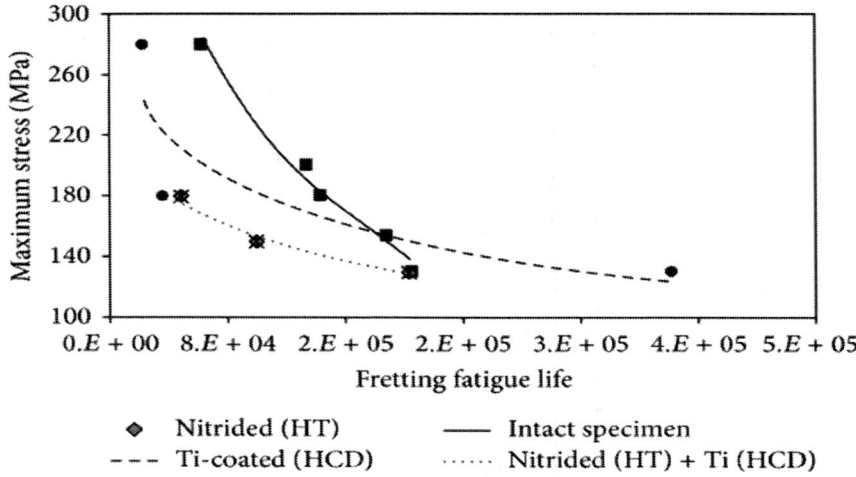

Figure 22: S-N curves for duplex surface treatment, (Ni (HT) + Ti (HCD)) [22].

Consequently, it is asserted that the duplex surface nitriding treatment at a low temperature followed by titanium coating by PVD magnetron sputtering technique generates the optimum AL7075-T6 performance in encountering fretting fatigue damage.

DISCUSSION

A multitude of surface modification techniques exists for decreasing fatigue damage. Significantly different mechanisms are reportedly capable of attaining this goal. At least, five diverse methods of increasing fretting fatigue resistance are applying compressive residual stress, reducing the friction coefficient, enhancing surface hardness, changing the surface chemistry, and increasing surface roughness. Table 3 tabulates a number of typical surface modifications applicable to Al7075-T6 alloy accompanied by their effects on fretting phenomena.

Table 3: Effect of surface modification techniques on fretting damage

Surface modification method	Decrease the coefficient of friction	Introduce compressive stress	Increase in hardness	Increase in surface roughness	Durability or adhesion	Economy	Mitigation of fretting wear	Mitigation of fretting fatigue
Nitriding	√	√√	√√	D	√√	√	√	√
Electroplating (Cr, Ni, etc.)	D	××	√	D	×	√	√	××
Hard anodizing	D	×	√	D	√	—	√	D
Shot peening	×	√	√	√√	—	√√	√√	√√
Laser shock peening	D	D	D	√	××	√	D	√
Deep rolling	√	√√	√	D	√√	√√	√	√√
IBED hard film	√	√	√	××	√	××	√√	√√
PVD hard coating	√	D	√	×	D	×	√	D

D: depending on condition; ×: bad effect; √: good effect; √√: very good effect, and ××: very bad effect.

Introducing compressive residual stresses into surface layers by means of surface modification is possibly among the most substantial mechanisms of decreasing fretting damage and consequently fretting fatigue [112]. Compressive stress causes fretting fatigue crack surfaces to close up, preventing crack propagation. Not only that, but compressive stress also decreases the tensile stress of fretting, subsequently lowering the rates of wear and crack propagation [113].

Selective deposition technique and surface treatments, such as ion implantation, shot peening, nitriding, and ion beam enhanced deposition, could induce compressive residual stress. Hence, such techniques are highly recommended to obtain better fretting resistance, mainly to improve the fretting fatigue strength. It is important to be noted that some types of surface modification including laser treatments and electroplating Cr are not proper to be applied for fretting fatigue life improvement as tensile residual stresses are often generated [114, 115].

The coefficient of friction plays an important role in fretting damage since greater friction force leads to higher strain fatigue or shear stress on surfaces and at interfaces. As such, delamination cracks or intensified fatigue failure could ensue. Lower friction coefficients can enhance the fretting fatigue strength in response to the minimized alternating tensile shear stresses. Elevated alternating stresses are a cause of high local strain fatigue and rapid fatigue crack initiation. Another concern is surface hardness, whereby a dramatic increase in surface hardness is followed by high residual tensile stress and reduced toughness in the surface layer, all of which are detrimental to fretting fatigue strength.

Changing the chemistry of the surface also possibly influences fretting damage. The fundamental function of oxidation during fretting and the important influence of oxide debris are understood by many scientists. Altering the surface chemistry by nitride, oxide, or carbide layer formation may help to enhance fretting resistance by having a lubricating effect.

The surface roughness influences on fretting resistance are fairly intricate and inconsistent. Good surface finishing emphasizes fretting damage, and to minimize such damage increasing surface roughness using different treatment techniques like shot peening is sometimes helpful. In some cases, increasing surface roughness may be the cause of a rising friction coefficient, which is not proper for fretting fatigue resistance. A rough surface has the potential to elevate stress, something very dangerous, especially under fatigue conditions [116–124].

Considering surface treatment means of mitigating fretting damage, specifically for Al 7075-T6, seven techniques have been drawn and discussed in this study. These particular methods are in great demand due to the positive effect on the fretting fatigue life of this aluminum alloy [125–129].

Deep rolling is absolutely successful in increasing fretting fatigue life at high cyclic load. The rolling force affects fatigue life in elevated rolling forces to enhance fatigue life by more than two times. It is thus concluded that by increasing rolling force an improvement in fatigue life is achievable. Deep rolling is also known to maintain low surface roughness. Compared to other techniques, deep rolling creates deeper compressive residual stress layers. Nevertheless, rolling force has the drawback of necessitating optimization and control.

Shot peening also follows the deep rolling's theory. They both produce a compressive residual stress layer. The distribution and depth of residual stress are affected by a variety of parameters like shot material and diameter, duration, target material, intensity of exposure, and shot velocity.

At low cycle fatigue (LCF) tests, SP enhances the material's fretting fatigue behavior much better than DR. SP increased fretting fatigue life by 300% for the tested specimens. The effect of DR on fretting fatigue resistance was more pronounced in high cycle fatigue, with a recorded improvement of about 700% at higher rolling forces.

A study of reshot peening techniques shows that the fretting fatigue lives of Al7075-T6 alloy are enhanced significantly. The fretting fatigue life of Al7075-T6 improved by 600% after 5 reshot peenings. It indicates that reshot peening has superior effect than simple shot peening.

Like shot peening and deep rolling, the mitigation of fretting fatigue life of specimens by using laser shock peening depends on the compressive residual stress layer. The deeper layer penetration can be achieved by LSP compared to SP. After laser shot peening treatment, significant improvement in fretting fatigue resistance can be obtained compared to other similar techniques.

Among the heat treatment methods meant to enhance the fretting fatigue of Al alloy 7075-T6, nitriding has always been an excellent option. In some cases, however, the results show that the fretting fatigue life of specimens with no surface modification is much higher than nitrided specimens.

PVD metal coatings including silver and gold were traditionally used to avoid fretting failure. The augmented fatigue life with PVD comes as a result of mainly compressive residual stresses during coating, in some cases at magnitudes of 7.08 GPa, and the excellent adhesion of coating to substrate.

In surface treatments by hard thin coating techniques, it is useful to be noted that fretting has a deleterious effect on the fatigue life of AL7075-T6 in uncoated and coated conditions at any bending stress [130].

Consequently, fretting can reduce the fatigue life of uncoated Al7075-T6 by about 30% in low cycle regions and by 57% in high cycle regions. Magnetron sputtering PVD coating in combination with hard anodizing substantially enhanced the fretting fatigue of AL7075-T6. As opposed to uncoated samples, a 119.55% enhancement in surface hardness of hard-anodized Al7075-T6 with a hardness of 393 HV was gained. At low bending stress, the fretting fatigue life of Al7075-T6 alloy is improved by the hard-anodized coating with hardness of 360 HV [131–134].

IBED technique demonstrates better film properties than similar films produced by PVD.

The superior ability of nitriding to enhance fretting fatigue is more evident when this technique is paired with magnetron sputtering PVD. Different studies have reported that none of nitriding and titanium coating could suitably enhance the fretting fatigue life of Al7075-T6. Hence, investigations showed the dramatic development in the fretting fatigue behavior of Al-7075-T6 when subjected to duplex surface treatment including nitriding followed by titanium coating. The duplex treatment is much more effective at lower stress. At a stress of 130 MPa, the fretting fatigue life improved by 100% but only by 13% at 280 MPa.

Accordingly, it is recognized that duplex surface treatment entailing nitriding at low temperature followed by titanium coating using magnetron-sputtering PVD technique results in the greatest Al7075-T6 performance against fretting fatigue damage.

SUMMARY

The effects of various methods including deep rolling, shot peening, laser shock peening, thin film hard coating using physical vapor deposition (PVD), hard anodizing, ion-beam-enhanced deposition (IBED), and nitriding to diminish the effect of fretting on the fatigue life of Al7075-T6 alloy are studied in detail. The following conclusions are derived.

- The deep rolling technique effectively improves the fretting fatigue behavior at high pressure. Increasing the rolling force increases fretting fatigue accordingly. DR maintains low surface roughness and great depth of compressive residual stress. The effect of DR on fretting fatigue resistance is deeper with high cycle fatigue, with up to 700% recorded. For low cycle fatigue (less than 10^5 cycles), SP improves fretting fatigue more effectively than DR, by 300%. Reshot peening technique seems to be much better than simple shot peening. After 5 reshot peenings, the fretting fatigue life of Al7075-T6 is improved by 600%.

- The fretting fatigue life for nitrided specimens with no surface modification is significantly lower than the intact samples. However, the duplex surface treatment comprising nitriding at a low temperature plus titanium coating using magnetron sputtering technique improved the fretting fatigue of Al 7075-T6 up to 100% at low stress. This technique could not sustain the same outcome at high stresses such as 280 MPa where fretting fatigue rose by merely 13%.

- Thin film hard coating is found to be a good choice for enhancing the fretting fatigue life of Al7075-T6. Fretting decreased the fatigue life of uncoated Al7075-T6 by 57% in high cycle regions and 30% in low cycle regions. The fretting fatigue lives of TiN-coated specimens are improved at high cyclic fatigue by 39% and 77% and at low cyclic fatigue by 61% and 16%, respectively, compared to the uncoated specimens.

- IBED technique demonstrates superior film properties to similar films prepared by PVD technique. Titanium coating via IBED reportedly increased the fatigue life of Al alloy 7075-T6 by 100% at low working stresses.

ACKNOWLEDGMENTS

The authors acknowledge the financial support under the University Malaya Research Grant (Grant no. UM.TNC2/RC/AET/GERAN (UMRG) RG133/11AET) from the University of Malaya, Malaysia. This research was partially funded by the University of Malaya under UMRG Programme Grant no. UM.TNC2/RC/AET/261/1/1/RP017-2012C.

REFERENCES

1. R. B. Waterhouse, Fretting Corrosion, Pergamon Press, 1972.

2. Y. Berthier, L. Vincent, and M. Godet, "Fretting fatigue and fretting wear," Tribology International, vol. 22, no. 4, pp. 235–242, 1989.

3. Y. Berthier, C. Colombie, L. Vicent, and M. Godet, "Fretting wear mechanisms and their effects on fretting fatigue," Journal of Tribology, vol. 110, no. 3, pp. 517–524, 1988.

4. P. Forsyth, "Occurrence of fretting fatigue failures in practice," in Fretting Fatigue, pp. 99–125, 1981.

5. D. A. Hills, "Mechanics of fretting fatigue," Wear, vol. 175, no. 1-2, pp. 107–113, 1994.

6. R. B. Waterhouse, Fretting Fatigue, IDEAS International, 1981.

7. R. B. Waterhouse and T. Lindley, "Fretting fatigue," in Proceedings of the International Symposium on Fretting Fatigue, Mechanical Engineering Publications, University of Sheffield, 1994.

8. D. Taylor and R. Waterhouse, "Wear, fretting and fretting fatigue," in Metal Behavior and Surface Engineering, pp. 13–35, 1989.

9. T. Juuma, "Torsional fretting fatigue strength of a shrink-fitted shaft with a grooved hub," Tribology International, vol. 33, no. 8, pp. 537–543, 2000.

10. J. Sato, "Recent trend in studies of fretting wear," Journal of Japan Society of Lubrication Engineers, vol. 30, no. 12, pp. 853–858, 1985.

11. G. H. Majzoobi, J. Nemati, A. J. Novin Rooz, and G. H. Farrahi, "Modification of fretting fatigue behavior of AL7075-T6 alloy by the application of titanium coating using IBED technique and

shot peening," Tribology International, vol. 42, no. 1, pp. 121–129, 2009.

12.   J. Sato, M. Shima, T. Sugawara, and A. Tahara, "Effect of lubricants on fretting wear of steel," Wear, vol. 125, no. 1-2, pp. 83–95, 1988.

13.   S. C. Gordelier and T. C. Chivers, "A literature review of palliatives for fretting fatigue," Wear, vol. 56, no. 1, pp. 177–190, 1979.

14.   Y. Qiu and B. J. Roylance, "The effect of lubricant additives on fretting wear," Lubrication Engineering, vol. 48, no. 10, pp. 801–808, 1992.

15.   R. Waterhouse, "The effect of surface treatment on the fatigue and fretting-fatigue of metallic materials," in Metal Treatments against Wear, Corrosion, Fretting and Fatigue, pp. 31–40, Pergamon Press, 1988.

16.   L. Xue, A. Koul, M. Bibby, W. Wallace, and M. Islam, "A survey of surface treatments to improve the fretting fatigue resistance of Ti-6Al-4V," International Journal of Fatigue, vol. 18, pp. 510–510, 1996.

17.   S. J. Harris, M. P. Overs, and A. J. Gould, "The use of coatings to control fretting wear at ambient and elevated temperatures," Wear, vol. 106, no. 1–3, pp. 35–52, 1985.

18.   R. C. Bill, "Fretting of AISI 9310 steel and selected fretting-resistant surface treatments," ASLE Transactions, vol. 21, no. 3, pp. 236–242, 1978.

19.   Y. Fu, J. Wei, and A. W. Batchelor, "Some considerations on the mitigation of fretting damage by the application of surface-modification technologies," Journal of Materials Processing Technology, vol. 99, no. 1, pp. 231–245, 2000.

20.   Y. Fu, N. L. Loh, A. W. Batchelor, D. Liu, J. He, and K. Xu, "Improvement in fretting wear and fatigue resistance of Ti-6Al-4V by application of several surface treatments and coatings," Surface and Coatings Technology, vol. 106, no. 2-3, pp. 193–197, 1998.

21.   L. Xue, M. Islam, A. K. Koul, M. Bibby, and W. Wallace, "Laser gas nitriding of Ti-6Al-4V. Part 1: optimization of the process," Advanced Performance Materials, vol. 4, no. 1, pp. 25–47, 1997.

22. G. H. Majzoobi and M. Jaleh, "Duplex surface treatments on AL7075-T6 alloy against fretting fatigue behavior by application of titanium coating plus nitriding," Materials Science and Engineering A, vol. 452-453, pp. 673–681, 2007.

23. D. Nowell and D. A. Hills, "Mechanics of fretting fatigue tests," International Journal of Mechanical Sciences, vol. 29, no. 5, pp. 355–365, 1987.

24. I. Nikitin and I. Altenberger, "Comparison of the fatigue behavior and residual stress stability of laser-shock peened and deep rolled austenitic stainless steel AISI 304 in the temperature range 25–600°C,"Materials Science and Engineering A, vol. 465, no. 1-2, pp. 176–182, 2007.

25. R. K. Nalla, I. Altenberger, U. Noster, G. Y. Liu, B. Scholtes, and R. O. Ritchie, "On the influence of mechanical surface treatments-deep rolling and laser shock peening-on the fatigue behavior of Ti-6Al-4V at ambient and elevated temperatures," Materials Science and Engineering A, vol. 355, no. 1-2, pp. 216–230, 2003.

26. T. C. Lindley, "Fretting fatigue in engineering alloys," International Journal of Fatigue, vol. 19, no. 1, pp. S39–S49, 1997.

27. M. A. Matin, W. P. Vellinga, and M. G. D. Geers, "Thermomechanical fatigue damage evolution in SAC solder joints," Materials Science and Engineering A, vol. 445-446, pp. 73–85, 2007.

28. M. Hirata, M. Maejima, K. Saruwatari, H. Shigeno, and M. Takaya, "Rotational bending fatigue of anodized coating of aluminum," Journal of the Surface Finishing Society of Japan, vol. 47, no. 4, pp. 376–377, 1996.

29. A. M. Cree, G. W. Weidmann, and R. Hermann, "Film-assisted fatigue crack propagation in anodized aluminium alloys," Journal of Materials Science Letters, vol. 14, no. 21, pp. 1505–1507, 1995.

30. P. Reybet Degat, Z. R. Zhou, and L. Vincent, "Fretting cracking behaviour on pre-stressed aluminium alloy specimens," Tribology International, vol. 30, no. 3, pp. 215–223, 1997.

31. I. R. McColl, S. J. Harris, Q. Hu, G. J. Spurr, and P. A. Wood, "Influence of surface and heat treatment on the fretting wear of an aluminium alloy reinforced with SiC particles," Wear, vol. 203-204, pp. 507–515, 1997.

32. R. L. Barrie, T. P. Gabb, J. Telesman et al., "Effectiveness of shot peening in suppressing fatigue cracking at non-metallic inclusions in Udimet® 720," Materials Science and Engineering A, vol. 474, no. 1-2, pp. 71–81, 2008.

33. N. Godja, N. Kiss, C. Löcker et al., "Preparation and characterization of spark-anodized Al-alloys: physical, chemical and tribological properties," Tribology International, vol. 43, no. 7, pp. 1253–1261, 2010.

34. R. Sadeler, S. Atasoy, A. Arlcl, and Y. Totik, "The fretting fatigue of commercial hard anodized aluminum alloy," Journal of Materials Engineering and Performance, vol. 18, no. 9, pp. 1280–1284, 2009.·

35. R. H. Oskouei and R. N. Ibrahim, "The effect of clamping compressive stresses on the fatigue life of Al 7075-T6 bolted plates at different temperatures," Materials and Design, vol. 34, pp. 90–97, 2012.

36. R. H. Oskouei and R. N. Ibrahim, "The effect of typical flight temperatures on the fatigue behaviour of Al 7075-T6 clamped plates," Materials Science and Engineering A, vol. 528, no. 3, pp. 1527–1533, 2011.·

37. E. Zalnezhad, A. A. D. Sarhan, and M. Hamdi, "Optimizing the PVD TiN thin film coating›s parameters on aerospace AL7075-T6 alloy for higher coating hardness and adhesion with better tribological properties of the coating surface," International Journal of Advanced Manufacturing Technology, vol. 64, no. 1–4, pp. 281–290, 2013.

38. G. H. Majzoobi, K. Azadikhah, and J. Nemati, "The effects of deep rolling and shot peening on fretting fatigue resistance of Aluminum-7075-T6," Materials Science and Engineering A, vol. 516, no. 1-2, pp. 235–247, 2009.

39. P. O›Hara, "Superfinishing and shot peening of surfaces to optimise roughness and stress," inProceedings of the International Conference on Computer Methods and Experimental Measurments for Surface Treatment Effects, pp. 321–330, 1999.

40. M. Kocan, A. Ostertag, and L. Wagner, "Shot peening and roller-burnishing to improve fatigue resistance of the (a+ß) titanium alloy Ti-6Al-4V," in Shot Peening, pp. 461–467, 2003.

41. X. P. Jiang, C.-S. Man, M. J. Shepard, and T. Zhai, "Effects of shot-peening and re-shot-peening on four-point bend fatigue behavior of Ti-6Al-4V," Materials Science and Engineering A, vol. 468-470, pp. 137–143, 2007.

42. R. Raghavan, R. Ayer, H. W. Jin, C. N. Marzinsky, and U. Ramamurty, "Effect of shot peening on the fatigue life of a Zr-based bulk metallic glass," Scripta Materialia, vol. 59, no. 2, pp. 167–170, 2008.

43. A. Ali, X. An, C. A. Rodopoulos et al., "The effect of controlled shot peening on the fatigue behaviour of 2024-T3 aluminium friction stir welds," International Journal of Fatigue, vol. 29, no. 8, pp. 1531–1545, 2007.

44. L. Wagner and G. Lütjering, "Influence of shot peening parameters on the surface layer properties and the fatigue life of Ti-6Al-4V," in Proceedings of the Second International Conference on Shot Peening (ICSP ‹48), pp. 194–200, 1984.

45. U. Martin, I. Altenberger, B. Scholtes, K. Kremmer, and H. Oettel, "Cyclic deformation and near surface microstructures of normalized shot peened steel SAE 1045," Materials Science and Engineering A, vol. 246, no. 1-2, pp. 69–80, 1998.

46. A. Turnbull, E. R. De Los Rios, R. B. Tait, C. Laurant, and J. S. Boabaid, "Improving the fatigue crack resistance of Waspaloy by shot peening," Fatigue and Fracture of Engineering Materials and Structures, vol. 21, no. 12, pp. 1513–1524, 1998.

47. B. P. Fairand, B. A. Wilcox, W. J. Gallagher, and D. N. Williams, "Laser shock-induced microstructural and mechanical property changes in 7075 aluminum," Journal of Applied Physics, vol. 43, no. 9, pp. 3893–3895, 1972.

48. A. H. Clauer, B. P. Fairand, and B. A. Wilcox, "Pulsed laser induced deformation in an Fe-3 Wt Pct Si alloy," Metallurgical Transactions A, vol. 8, no. 1, pp. 119–125, 1977.

49. A. H. Clauer, B. P. Fairand, and B. A. Wilcox, "Laser shock hardening of weld zones in aluminum alloys," Metallurgical Transactions A, vol. 8, pp. 1871–1876, 1977.

50. B. P. Fairand and A. H. Clauer, "Laser generation of high-amplitude stress waves in materials," Journal of Applied Physics, vol. 50, no. 3, pp. 1497–1502, 1979.

51. A. Clauer, B. Fairand, and E. Metzbower, Applications of Lasers in Materials Processing, American Society for Metals, Metals Park, Ohio, USA, 1979.

52. A. King, A. Steuwer, C. Woodward, and P. J. Withers, "Effects of fatigue and fretting on residual stresses introduced by laser shock peening," Materials Science and Engineering A, vol. 435-436, pp. 12–18, 2006.·

53. P. Forget, J. L. Strude, M. Jeandin, J. Lu, and L. Castex, "Laser shock surface treatment of Ni-based superalloys," Materials and Manufacturing Processes, vol. 5, no. 4, pp. 501–528, 1990.

54. P. Peyre and R. Fabbro, "Laser shock processing: a review of the physics and applications," Optical and Quantum Electronics, vol. 27, no. 12, pp. 1213–1229, 1995.

55. J. P. Chu, J. M. Rigsbee, G. Banas, F. V. Lawrence, and H. E. Elsayed-Ali, "Effects of laser-shock processing on the microstructure and surface mechanical properties of Hadfield manganese steel,"Metallurgical and Materials Transactions A: Physical Metallurgy and Materials Science, vol. 26, no. 6, pp. 1507–1517, 1995.

56. P. Peyre, R. Fabbro, P. Merrien, and H. P. Lieurade, "Laser shock processing of aluminium alloys. Application to high cycle fatigue behaviour," Materials Science and Engineering A, vol. 210, no. 1-2, pp. 102–113, 1996.

57. C. B. Dane, L. A. Hackel, J. Daly, and J. Harrison, "Shot peening with lasers," Advanced Materials and Processes, vol. 153, no. 5, pp. 37–38, 1998.

58. P. Peyre, X. Scherpereel, L. Berthe et al., "Surface modifications induced in 316L steel by laser peening and shot-peening. Influence on pitting corrosion resistance," Materials Science and Engineering A, vol. 280, no. 2, pp. 294–302, 2000.

59. J. P. Chu, J. M. Rigsbee, G. Bana , and H. E. Elsayed-Ali, "Laser-shock processing effects on surface microstructure and mechanical properties of low carbon steel," Materials Science and Engineering A, vol. 260, no. 1-2, pp. 260–268, 1999.

60. G. H. Majzoobi and A. R. Ahmadkhani, "The effects of multiple re-shot peening on fretting fatigue behavior of Al7075-T6," Surface and Coatings Technology, vol. 205, no. 1, pp. 102–109, 2010.

61. A. H. Clauer, "Laser shock peening for fatigue resistance," in Surface Performance of Titanium, J. K. Gregory, H. J. Rack, and D. Eylon, Eds., pp. 217–230, TMS, Warrendale, Pa, USA, 1996.

62. A. H. Clauer, J. H. Holbrook, and B. P. Fairand, "Effects of laser induced shock waves on metals," inShock Waves and High-Strain-Rate Phenomena in Metals, pp. 675–702, Springer, 1981.

63. S. F. Design and E. Committee, SAE Manual on Shot Peening (SAE HS-84), Society of Automotive Engineers, 1991.

64. M. Kobayashi, T. Matsui, and Y. Murakami, "Mechanism of creation of compressive residual stress by shot peening," International Journal of Fatigue, vol. 20, no. 5, pp. 351–357, 1998.

65. G. Hammersley, L. A. Hackel, and F. Harris, "Surface prestressing to improve fatigue strength of components by laser shot peening," Optics and Lasers in Engineering, vol. 34, no. 4–6, pp. 327–337, 2000.

66. P. S. Prevéy, M. J. Shepard, and P. R. Smith, "The effect of Low Plasticity Burnishing (LPB) on the HCF performance and FOD resistance of Ti-6Al-4V," DTIC Document, 2001.

67. W. Z. Zhuang and G. R. Halford, "Investigation of residual stress relaxation under cyclic load,"International Journal of Fatigue, vol. 23, no. 1, pp. S31–S37, 2001.

68. S. G. Harris, E. D. Doyle, A. C. Vlasveld, J. Audy, J. M. Long, and D. Quick, "Influence of chromium content on the dry machining performance of cathodic arc evaporated TiAlN coatings," Wear, vol. 254, no. 1-2, pp. 185–194, 2003.

69. C.-H. Hsu and Y.-D. Chen, "A study on the abrasive and erosive wear behavior of arc-deposited Cr-N-O coatings on tool steel," Thin Solid Films, vol. 517, no. 5, pp. 1655–1661, 2009.

70. O. Jin, S. Mall, J. H. Sanders, and S. K. Sharma, "Durability of Cu-Al coating on Ti-6Al-4V substrate under fretting fatigue," Surface and Coatings Technology, vol. 201, no. 3-4, pp. 1704–1710, 2006.

71. K. Schouterden, B. Blanpain, J. P. Celis, and O. Vingsbo, "Fretting of titanium nitride and diamond-like carbon coatings at high frequencies and low amplitude," Wear, vol. 181-183, no. 1, pp. 86–93, 1995.

72. T. Liskiewicz, S. Fouvry, and B. Wendler, "Impact of variable loading conditions on fretting wear,"Surface and Coatings Technology, vol. 163-164, pp. 465–471, 2003.

73. N. Ohmae, T. Nakai, and T. Tsukizoe, "Prevention of fretting by ion plated film," Wear, vol. 30, no. 3, pp. 299–309, 1974.

74. N. Ohmae, T. Tsukizoe, and T. Nakai, "Ion-plated thin films for anti-wear applications," Journal of Lubrication Technology, vol. 100, no. 1, pp. 129–135, 1978.

75. B. Subramanian and M. Jayachandran, "Characterization of reactive magnetron sputtered nanocrystalline titanium nitride (TiN) thin films with brush plated Ni interlayer," Journal of Applied Electrochemistry, vol. 37, no. 9, pp. 1069–1075, 2007.

76. J. H. Hsieh, C. Liang, C. H. Yu, and W. Wu, "Deposition and characterization of TiAlN and multi-layered TiN/TiAlN coatings using unbalanced magnetron sputtering," Surface and Coatings Technology, vol. 108-109, pp. 132–137, 1998.

77. J.-H. Huang, K.-W. Lau, and G.-P. Yu, "Effect of nitrogen flow rate on structure and properties of nanocrystalline TiN thin films produced by unbalanced magnetron sputtering," Surface and Coatings Technology, vol. 191, no. 1, pp. 17–24, 2005.

78. C. W. Tan and J. Miao, "Optimization of sputtered Cr/Au thin film for diaphragm-based MEMS applications," Thin Solid Films, vol. 517, no. 17, pp. 4921–4925, 2009.

79. B. Blanpain, H. Mohrbacher, E. Liu, J. P. Celis, and J. R. Roos, "Hard coatings under vibrational contact conditions," Surface and Coatings Technology, vol. 74-75, no. 2, pp. 953–958, 1995.

80. H. Ezuber, A. El-Houd, and F. El-Shawesh, "A study on the corrosion behavior of aluminum alloys in seawater," Materials and Design, vol. 29, no. 4, pp. 801–805, 2008.

81. A. Camargo and H. Voorwald, "Influence of anodization on the fatigue strength of 7050-T7451 aluminium alloy," Fatigue and Fracture of Engineering Materials and Structures, vol. 30, no. 11, pp. 993–1007, 2007.

82. B. Rajasekaran, S. Ganesh Sundara Raman, L. Rama Krishna, S. V. Joshi, and G. Sundararajan, "Influence of microarc oxidation and hard anodizing on plain fatigue and fretting fatigue behaviour of Al-Mg-Si alloy," Surface and Coatings Technology, vol. 202, no. 8, pp. 1462–1469, 2008.

83. A. A. D. Sarhan, E. Zalnezhad, and M. Hamdi, "The influence of higher surface hardness on fretting fatigue life of hard anodized aerospace AL7075-T6 alloy," Materials Science and Engineering A, 2012.

84. J. A. Ghani, I. A. Choudhury, and H. H. Hassan, "Application of Taguchi method in the optimization of end milling parameters," Journal of Materials Processing Technology, vol. 145, no. 1, pp. 84–92, 2004.

85. M. Farooq and Z. H. Lee, "Computations of the optical properties of metal/insulator-composites for solar selective absorbers," Renewable Energy, vol. 28, no. 9, pp. 1421–1431, 2003.

86. F. A. Smidt, "Use of ion beam assisted deposition to modify the microstructure and properties of thin films," International Materials Reviews, vol. 35, no. 2, pp. 61–128, 1990.

87. J. K. Hirvonen, "Ion beam assisted thin film deposition," Materials Science Reports, vol. 6, no. 6, pp. 215–274, 1991.

88. H. Fladry, N. Tegen, and G. K. Wolf, "Ion beam induced adhesion improvement of metal layers—a comparative study on composite layers," Nuclear Instruments and Methods in Physics Research Section B: Beam Interactions with Materials and Atoms, vol. 91, no. 1–4, pp. 575–579, 1994.

89. G. K. Hubler, "Fundamentals of ion-beam-assisted deposition: technique and film properties," Materials Science and Engineering A, vol. 115, pp. 181–192, 1989.

90. X.-M. He, W.-Z. Li, and H.-D. Li, "Ion beam assisted deposition of diamond-like carbon onto steel materials: preparation and advantages," Surface and Coatings Technology, vol. 84, no. 1–3, pp. 414–419, 1996.

91. K.-H. Bäther, U. Herrmann, and A. Schröer, "Ion-beam-assisted deposition of magnetron-sputtered metal nitrides," Surface and Coatings Technology, vol. 74-75, no. 2, pp. 793–801, 1995.

92. M. Zaytouni, J. P. Riviere, M. F. Denanot, and J. Allain, "Structural characterization of SiC films prepared by dynamic ion mixing," Thin Solid Films, vol. 287, no. 1-2, pp. 1–7, 1996.

93. G. K. Hubler and J. A. Sprague, "Energetic particles in PVD technology: particle-surface interaction processes and energy-particle relationships in thin film deposition," Surface and Coatings Technology, vol. 81, no. 1, pp. 29–35, 1996.

94. T. Bell, J. Lanagan, P. H. Morton, H. W. Bergmann, and A. M. Staines, "Surface engineering of titanium with nitrogen," Surface Engineering, vol. 2, no. 2, pp. 133–143, 1986.

95. B. S. Yilba , A. Z. Sahin, A. Z. Al-Garni et al., "Plasma nitriding of Ti-6Al-4V alloy to improve some tribological properties," Surface and Coatings Technology, vol. 80, no. 3, pp. 287–292, 1996.

96. S. Goudarzi, K. Khojier, H. Savaloni, and E. Zalnezhad, "On the dependence of mechanical and tribological properties of sputtered chromium nitride thin films on deposition power," Advanced Materials Research, vol. 829, pp. 352–356.

97. S. K. Wu, H. C. Lin, and C. Y. Lee, "Gas nitriding of an equiatomic TiNi shape memory alloy. II: hardness, wear and shape memory ability," Surface and Coatings Technology, vol. 113, no. 1-2, pp. 13–16, 1999.

98. K. C. Chen and G. J. Jaung, "D.c. diode ion nitriding behavior of titanium and Ti-6Al-4V," Thin Solid Films, vol. 303, no. 1-2, pp. 226–231, 1997.

99. T. M. Muraleedharan and E. I. Meletis, "Surface modification of pure titanium and Ti6A14V by intensified plasma ion nitriding," Thin Solid Films, vol. 221, no. 1-2, pp. 104–113, 1992.

100. L. Xue, M. U. Islam, A. K. Koul, W. Wallace, and M. Bibby, "Laser gas nitriding of Ti-6Al-4V alloy,"Materials and Manufacturing Processes, vol. 12, no. 5, pp. 799–817, 1997.

101. A. Zhecheva, W. Sha, S. Malinov, and A. Long, "Enhancing the microstructure and properties of titanium alloys through nitriding and other surface engineering methods," Surface and Coatings Technology, vol. 200, no. 7, pp. 2192–2207, 2005.

102. J. M. O›Brien and D. Goodman, "Plasma (ion) nitriding," in ASM Handbook, vol. 4, pp. 420–424, ASM International, 1991.

103. E. S. Puchi-Cabrera, F. Matínez, I. Herrera, J. A. Berríos, S. Dixit, and D. Bhat, "On the fatigue behavior of an AISI 316L stainless steel coated with a PVD TiN deposit," Surface and Coatings Technology, vol. 182, no. 2-3, pp. 276–286, 2004.

104. E. Zalnezhad, A. A. D. Sarhan, and M. Hamdi, "Investigating the fretting fatigue life of thin film titanium nitride coated aerospace Al7075-T6 alloy," Materials Science and Engineering A, vol. 559, pp. 436–446, 2013.

105. E. S. Puchi-Cabrera, M. H. Staia, J. Lesage et al., "Fatigue behavior of AA7075-T6 aluminum alloy coated with ZrN by PVD," International Journal of Fatigue, vol. 30, no. 7, pp. 1220–1230, 2008.

106. P. S. Pao, S. J. Gill, C. R. Feng, and K. K. Sankaran, "Corrosion-fatigue crack growth in friction stir welded Al 7050," Scripta Materialia, vol. 45, no. 5, pp. 605–612, 2001.

107. R. G. Rateick Jr., R. J. Griffith, D. A. Hall, and K. A. Thompson, "Relationship of microstructure to fatigue strength loss in anodised aluminium-copper alloys," Materials Science and Technology, vol. 21, no. 10, pp. 1227–1235, 2005.

108. A. Monsalve, M. Páez, M. Toledano, A. Artigas, Y. Sepúlveda, and N. Valencia, "S-N-P curves in 7075 T7351 and 2024 T3 aluminium alloys subjected to surface treatments," Fatigue and Fracture of Engineering Materials and Structures, vol. 30, no. 8, pp. 748–758, 2007.

109. Q. Zhang and W. Wang, "Study of anodizing behavior and corrosion resistance of 7050 T7451 alloy,"Materials Science and Engineering A, vol. 280, no. 1, pp. 168–172, 2000.

110. M. L. Sharp, G. E. Nordmark, and C. C. Menzemer, Fatigue Design of Aluminium Components and Structures, McGraw-Hill, New York, NY, USA, 1996.

111. Y. Fu, N. Lam Loh, A. W. Batchelor, X. Zhu, K. Xu, and J. He, "Preparation and fretting wear behavior of ion-beam-enhanced-deposition CrN films," Materials Science and Engineering A, vol. 265, no. 1-2, pp. 224–232, 1999.

112. R. B. Waterhouse and A. J. Trowsdale, "Residual stress and surface roughness in fretting fatigue,"Journal of Physics D: Applied Physics, vol. 25, no. 1, pp. A236–A239, 1992.

113. G. Leadbeater, B. Noble, and R. Waterhouse, "The fatigue of an aluminium alloy produced by fretting on a shot peened surface," Advances in Fracture Research, vol. 3, pp. 2125–2132, 1986.

114. A. J. Gould, P. J. Boden, and S. J. Harris, "Phosphorus distribution in electroless nickel deposits," Surface Technology, vol. 12, no. 1, pp. 93–102, 1981.

115. H. Gao, H. Gu, and H. Zhou, "Sliding wear and fretting fatigue resistance of amorphous Ni-P coatings,"Wear, vol. 142, no. 2, pp. 291–301, 1991.

116. U. Bryggman and S. Söderberg, "Contact conditions and surface degradation mechanisms in low amplitude fretting," Wear, vol. 125, no. 1-2, pp. 39–52, 1988.

117. E. Zalnezhad, A. A. D. M. Sarhan, and M. Hamdi, "Prediction of TiN coating adhesion strength on aerospace AL7075-T6 alloy using fuzzy rule based system," International Journal of Precision Engineering and Manufacturing, vol. 13, no. 8, pp. 1453–1459, 2012.

118. G. H. Majzoobi, R. Hojjati, M. Nematian, E. Zalnejad, A. R. Ahmadkhani, and E. Hanifepoor, "A new device for fretting fatigue testing," Transactions of the Indian Institute of Metals, vol. 63, no. 2-3, pp. 493–497, 2010.

119. E. Zalnezhad, A. A. D. M. Sarhan, and M. Hamdi, "Surface hardness prediction of CrN thin film coating on AL7075-T6 alloy using fuzzy logic system," International Journal of Precision Engineering and Manufacturing, vol. 14, no. 3, pp. 467–473, 2013.

120. E. Mohseni, E. Zalnezhad, and A. R. Bushroa, "Comparative investigation on the adhesion of hydroxyapatite coating on Ti-6Al-4V implant: a review paper," International Journal of Adhesion and Adhesives, vol. 48, pp. 238–257, 2014.

121. E. Zalnezhad, A. A. D. Sarhan, and M. Hamdi, "A fuzzy logic based model to predict surface hardness of thin film TiN coating on aerospace AL7075-T6 alloy," International Journal of Advanced Manufacturing Technology, vol. 68, pp. 415–423, 2013.

122. M. S. Farhan, E. Zalnezhad, A. R. Bushroa, and A. A. D. Sarhan, "Electrical and optical properties of indium-tin oxide (ITO) films by ion-assisted deposition (IAD) at room temperature," International Journal of Precision Engineering and Manufacturing, vol. 14, no. 8, pp. 1465–1469, 2013.

123. S. Baradaran, W. J. Basirun, E. Zalnezhad, M. Hamdi, A. A. D. Sarhan, and Y. Alias, "Fabrication and deformation behaviour of multilayer $Al_2O_3/Ti/TiO_2$ nanotube arrays," Journal of the Mechanical Behavior of Biomedical Materials, vol. 20, pp. 272–282, 2013.

124. A. A. D. Sarhan, E. Zalnezhad, and M. Hamdi, "The influence of higher surface hardness on fretting fatigue life of hard anodized aerospace AL7075-T6 alloy," Materials Science and Engineering A, vol. 1, pp. 23–40, 2012.

125. E. Zalnezhad, A. A. D. Sarhan, and M. Hamdi, "A fuzzy logic based model to predict surface hardness of thin film TiN coating on aerospace AL7075-T6 alloy," International Journal of Advanced Manufacturing Technology, vol. 49, pp. 256–265, 2014.

126. E. Zalnezhad, S. Baradaran, A. R. Bushroa, and A. A. D. Sarhan, "Mechanical property enhancement of Ti-6Al-4V by multilayer thin solid film Ti/TiO$_2$ nanotubular array coating for biomedical application,"Metallurgical and Materials Transactions A, vol. 45, pp. 785–797, 2014.

127. S. Goudarzi, K. Khojier, H. Savaloni, and E. Zalnezhad, "On the dependence of mechanical and tribological properties of sputtered chromium nitride thin films on deposition power," Advanced Materials Research, vol. 829, pp. 352–356, 2014.

128. M. S. Farhan, E. Zalnezhad, and A. R. Bushroa, "Investigation of optical and structural properties of ion-assisted deposition (IAD) ZrO$_2$ thin films," International Journal of Precision Engineering and Manufacturing, vol. 14, pp. 1997–2002, 2013.

129. E. Zalnezhad, A. A. D. Sarhan, and M. Hamdi, "Fretting fatigue life evaluation of multilayer Cr-CrN-coated Al7075-T6 with higher adhesion strength-fuzzy logic approach," International Journal of Advanced Manufacturing Technology, vol. 69, pp. 1153–1164, 2013.

130. M. S. Farhan, E. Zalnezhad, and A. R. Bushroa, "Properties of Ta$_2$O$_5$ thin films prepared by ion-assisted deposition," Materials Research Bulletin, vol. 48, no. 10, pp. 4206–4209, 2013.

131. E. Zalnezhad, A. A. D. Sarhan, and M. Hamdi, "Investigating the effects of hard anodizing parameters on surface hardness of hard anodized aerospace AL7075-T6 alloy using fuzzy logic approach for fretting fatigue application," International Journal of Advanced Manufacturing Technology, vol. 68, pp. 453–464, 2013. · ·

132. E. Zalnezhad, A. A. D. Sarhan, and P. Jahanshahi, "A new fretting fatigue testing machine design, utilizing rotating-bending principle approach," The International Journal of Advanced Manufacturing Technology, vol. 70, pp. 2211–2219, 2014.

133. E. Zalnezhad, A. A. D. Sarhan, and M. Hamdi, "Adhesion strength predicting of Cr/CrN coated Al7075 using fuzzy logic system for fretting fatigue life enhancement," in Proceedings of the World Congress on Engineering and Computer Science, vol. 1, pp. 2–8, 2013.

134. E. Zalnezhad and A. D. Ahmed Sarhan, "Multilayer thin film CrN coating on aerospace AL7075-T6 alloy for surface integrity enhancement," The International Journal of Advanced Manufacturing Technology, vol. 72, pp. 1491–1502, 2014.

Chapter 3

Recent Advances in the Study of Biocorrosion - An Overview

Iwona B. Beech[1] and Christine C. Gaylarde[2]

[1]University of Portsmouth, UK

[2]Microbiological Resources Center - MIRCEN, Departamento de Solos, Universidade Federal do Rio Grande do Sul - UFRGS. Porto Alegre, RS, Brasil

ABSTRACT

Biocorrosion processes at metal surfaces are associated with microorganisms, or the products of their metabolic activities including enzymes, exopolymers, organic and inorganic acids, as well as volatile compounds such as ammonia or hydrogen sulfide. These can affect cathodic and/or anodic reactions, thus altering electrochemistry at the biofilm/metal interface. Various mechanisms of biocorrosion, reflecting the variety of physiological activities carried out by different types of microorganisms, are identified and recent insights into these mechanisms reviewed. Many modern investigations have centered on the microbially-influenced corrosion of ferrous and copper alloys and particular microorganisms of interest have been the sulfate-reducing bacteria and metal (especially manganese)-depositing bacteria. The importance of microbial consortia and the role of extracellular polymeric substances in biocorrosion are emphasized. The contribution to the study of biocorrosion of modern analytical techniques, such as atomic force microscopy, Auger electron, X-ray photoelectron and Mössbauer spectroscopy, attenuated total reflectance Fourier transform infrared spectroscopy and microsensors, is discussed.

INTRODUCTION

In natural and man-made environments corrosion occurs when materials made of pure metals and/or their mixtures (alloys) undergo a chemical change from the ground state to an ionized species. Corrosion is an electrochemical process consisting of an anodic reaction involving the ionization (oxidation) of the metal (the corrosion reaction), and a cathodic reaction based on the reduction of a chemical species. Many textbooks cover basic corrosion concepts and may be consulted for further details (16, 99). These reactions can be influenced by microbial activities, especially when the organisms are in close contact with the metal surface forming a biofilm (Fig. 1). The resulting metal deterioration is known as biocorrosion, or microbially-influenced corrosion (MIC).

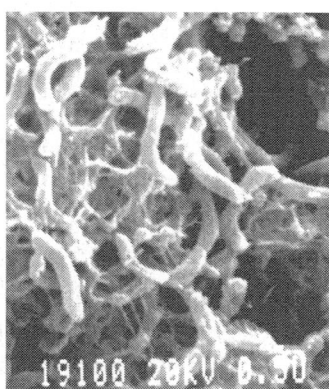

Figure 1: Biofilm formed by sulfate-reducing bacteria on the surface of mild steel, visualized using SEM.

Biofilms consist of microbial cells, their extracellular polymeric substances (EPS), which facilitate irreversible attachment of cells to the surface, inorganic precipitates derived from the bulk aqueous phase and/or corrosion products of the metal substratum. EPS consist of a complex mixture of cell-derived polysaccharides, proteins, lipids and nucleic acids. Microorganisms, and/or products of their metabolic activities, e.g. enzymes, exopolymers, organic and inorganic acids, as well as volatile compounds such as ammonia or hydrogen sulfide, can affect cathodic and/or anodic reactions at metal surfaces, thus altering electrochemical processes at the biofilm/metal interface. However, the number of attached microorganisms does not necessarily correlate with the extent of corrosion (6), a fact that has long been known for suspended cells (41). It is the metabolic status of the cells that is believed to be the relevant parameter, but to date no clear consensus has been reached linking specific bacterial metabolic rates to observed corrosion rates.

Economic Losses Caused by Biocorrosion

There are no official figures for the cost of MIC, but some indication of its importance can be gained from individual companies or sectors of industry.

Escom, the national power utility of South Africa that provides 90% of power requirements for the country, has detected MIC of carbon steel in cooling water systems in virtually all their power plants. The costs associated with repairs and down time are millions of dollars annually (14). Under-deposit pitting corrosion of heat exchanger tubing in nuclear power generating plants operated by Ontario Hydro of Canada has been estimated to cost the corporation $ 300,000 per unit per day in replacement energy costs (18). Corrosion problems have cost the nuclear utility billions of dollars in replacement costs alone (Jones, 1996). Losses in the oil and gas industry are also substantial; Jack et al. (50) estimated that 34% of the corrosion damage experienced by one oil company was related to microorganisms. In the 1950s, MIC-related costs of repair and replacement of piping material used in different types of service in the USA were estimated to be around $ 0.5-2 Billion per annum. Booth (15), in the UK, suggested that 50% of corrosion failures in pipelines involved MIC, while Flemming (40) proposed that approximately 20% of all corrosion damage to metallic materials is microbially influenced. Replacement costs for biocorroded gas mains in the UK were recently reported to be £250 Million per annum. Often, financial losses due to damage of equipment by biocorrosion are combined with those resulting from biofouling. While the two phenomena may be associated, they do not cause the same type of damage. The costs associated with MIC usually include the costs of prevention of both MIC and biofouling; since these are based on a limited understanding of the phenomena, they could be underestimated.

Mechanisms of Biocorrosion

MIC does not invoke any new electrochemical mechanisms of corrosion; rather, it is the result of a microbiologically-influenced change that promotes the establishment or maintenance of physico-chemical reactions not normally favoured under otherwise similar conditions. Various mechanisms of biocorrosion, which reflect the variety of physiological activities carried out by different types of microorganisms, have been identified; however, it must be remembered that, in nature, these microbial processes do not act in isolation, but in concert with the chemical and electrochemical forces in the particular environment.

Activities of Microorganisms as the Driving Force for Biocorrosion

Microorganisms implicated in biocorrosion of metals such as iron, copper and aluminium and their alloys are physiologically diverse. Their ability to influence the corrosion of many metals normally considered corrosion resistant, in a variety of environments, makes microorganisms a real threat to the stability of those metals.

The main types of bacteria associated with corrosion failures of cast iron, mild and stainless steel structures are sulfate-reducing bacteria (45), sulfur-oxidising bacteria (25), iron-oxidising/reducing bacteria (86 and references therein), manganese-oxidizing bacteria (30), and bacteria secreting organic acids and exopolymers or slime (25, 116). These organisms can coexist in naturally occurring biofilms, often forming synergistic communities (consortia) that are able to affect electrochemical processes through co-operative metabolism not seen in the individual species (34). Much recent research activity has centered on the role of "quorum sensing" molecules, such as acylhomoserine lactones, in control of microbial activities in biofilms (29, 83), with the aim of using this knowledge to reduce problematical biofilm formation in industry (115).

Sulfate-Reducing Bacteria (SRB)

SRB are a group of diverse anaerobes which carry out dissimilatory reduction of sulfur compounds such as sulfate, sulfite, and thiosulfate and even sulfur itself to sulfide (4, 74). Although SRB are often considered to be strictly anaerobic, some genera tolerate oxygen (1, 48) and at low dissolved oxygen concentrations certain SRB are able to respire with Fe^{3+} or even oxygen with hydrogen acting as electron donor (32, 94). Excellent reviews on the ecology and physiology of SRB are available in the literature (93, 117, Barton, 1995).

Oil, gas and shipping industries are seriously affected by the sulfides generated by SRB (46 and references therein). Biogenic sulfide production leads to health and safety problems, environmental hazards and severe economic losses due to reservoir souring (increased sulfur content) and the corrosion of equipment. Since the beginning of investigations into the effects of SRB on corrosion of cast iron in 1930s, the role of these bacteria in the pitting corrosion of various metals and their alloys in both aquatic and terrestrial environments, under anoxic as well as oxygenated conditions, has been confirmed. Several models have been proposed to explain the mechanisms by which SRB can influence the corrosion of steel (Table 1) and it is clear that sulfate reducing activity is in some way involved. The product of this activity, sulfide, is corrosive; however, chemically-derived sulfide does not have the same degree of aggressivity (73, 79, 105), demonstrating the importance of bioprocesses and the irrelevance of experiments using abiotic, as opposed to biologically derived compounds. Videla et al. (107) used energy dispersion X-ray analysis, X-ray photoelectron spectroscopy, X-ray diffraction, electron microprobe analysis, scanning electron microscopy and atomic force microscopy to demonstrate that the composition and structure of the sulfide films formed on carbon steel in the presence of the SRB, Desulfovibrio alaskensis, (biotic sulfides) were different from those formed in sterile, sulfide-containing medium (abiotic sulfides). Recent reviews clearly state that one predominant mechanism may not exist in SRB-influenced corrosion and that a number of factors are involved (47, 60).

Table 1: Suggested mechanisms of metal corrosion by SRB

Corrosive process/substance	Reference(s)
Cathodic depolarization* by hydrogenase	Von Wolzogen Kühr and van der Vlugt, 1934; Bryant et al., 1991.
Anodic depolarization*	Salvarezza and Videla, 1984; Daumas et al., 1988; Crolet, 1992.
Sulfide	Little et al., 1998.
Iron sulfides	King and Wakerley, 1973.
A volatile phosphorus compound	Iverson and Ohlson, 1983.
Fe-binding exopolymers	Beech and Cheung, 1995; Beech et al., 1996, 1998, 1999.
Sulfide-induced stress corrosion cracking	Edyvean et al., 1998.
Hydrogen-induced cracking or blistering	Edyvean et al., 1998.

*depolarization is an acceleration of the corrosion reaction and may involve removal of cathodic or anodic reactants.

Considerable work has centered on the influence of ferrous ions on SRB action on steel alloys. Obuekwe et al. (86) reported extensive pitting of mild steel when ferrous and sulfide ions were being formed concurrently. When only sulfide was produced, corrosion rates first increased and then declined due to the formation of a protective FeS film. High levels of soluble iron prevented the formation of such protective layers. Moulin et al. (84) demonstrated that high soluble iron levels could lead to high corrosion rates of piling grade carbon steel and Gubner et al. (44) showed that this was linked to a decrease in pH. The hydrogenase of Desulfovibrio vulgaris (Hildenborough) has been shown to be regulated by Fe^{2+} availability (20), offering yet another mechanism whereby corrosion may be affected, as assessed by Cheung and Beech (23). Thus the influence of iron ions on SRB-influenced corrosion is a complex phenomenon; this was reviewed by Videla et al. (108).

The impact of sulfides on the corrosion of copper alloys has recently received considerable attention. Copper alloys are attacked after only one day in seawater containing 0.01 ppm sulfide. In the presence of sulfide ions, an interstitial cuprous sulfide compound, with the general stoichiometry $Cu_{2-x}S$ ($0 < x < 1$), is formed; copper ions migrate through this layer and react with more sulfide. The result can be the production of thick scale (71).

Specific removal of nickel from 90-10 and 70-30 Cu-Ni has been reported in seawater containing SRB (64, 112). Spalling of the nickel-enriched region of the metal occurs during exposure to flowing seawater, exposing fresh metal and causing further dissolution of the alloy. Welds also exhibit this type of corrosion in the presence of SRB (63).

SRB can induce corrosion of zinc and lead based alloys. The corrosion product on zinc is reported to be sphalerite (ZnS), while the action of SRB on lead carbonates produces galena (PbS), also found as a corrosion product on lead-tin alloys (71).

Metal-Reducing Bacteria (MRB)

Microorganisms are known to promote corrosion of iron and its alloys through reactions leading to the dissolution of corrosion-resistant oxide films on the metal surface. This results in the protective passive layers on e.g. stainless steel surfaces being lost or replaced by less stable reduced metal films that allow further corrosion to occur. Despite its widespread occurrence in nature and likely importance to industrial corrosion, bacterial metal reduction has not been seriously considered in corrosion reactions until recently.

Numerous types of bacteria, including those from the genera Pseudomonas (86) and Shewanella (85) are able to carry out manganese and/or iron oxide reduction and have been shown to influence corrosion reactions. It has been demonstrated that in cultures of Shewanella putrefaciens, iron oxide-surface contact was required for bacterial cells to mediate reduction of these metals (85). The rate of reaction depended on the type of oxide film under attack (69).

Metal-Depositing Bacteria (MDB)

Bacteria of the genera *Siderocapsa, Gallionella, Leptothrix, Sphaerotilus, Crenothrix* and *Clonothrix* participate in the biotransformation of oxides of metals such as iron and manganese (43). Iron-depositing bacteria (*e.g.,Gallionella* and *Leptothrix*) oxidize Fe^{2+}, either dissolved in the bulk medium or precipitated on a surface, to Fe^{3+}. Bacteria of the genera given above are also capable of oxidizing manganous ions to manganic ions with concomitant deposition of manganese dioxide (70).

A role in the corrosion of steels has been proposed for sheathed filamentous bacteria detected by microscopy in naturally formed corrosion deposits (57, 75, and 104). These bacteria have been typically associated with formation of tubercles (macroscopic deposits containing microorganisms, inorganic and organic materials) and consequent under-deposit pitting attack on stainless steel. The corrosion resistance of alloys such as stainless steels is due to the formation of a thin passive oxide film. The formation of organic and inorganic deposits by MDB on the oxide surface compromises the stability of this film. Dense accumulations of MDB on the metal surface may thus promote corrosion reactions by the deposition of cathodically-reactive ferric and manganic oxides and the local consumption of oxygen by bacterial respiration in the deposit. However, care must be taken in considering microorganisms in corrosion products to be the causal agent. Some bacteria are known to adhere preferentially to corrosion products and thus will be present in high numbers even when playing no role in the primary corrosion process (72).

MDB have been shown to promote ennoblement of metals (a change to more positive values of pitting potential) and pitting corrosion. It has been demonstrated that the formation of a surface biofilm containing the sheath-forming, manganese-depositing bacterium, *Leptothrix discophora,* resulted in the ennoblement of 316L stainless steel (31). The biofilm was proposed to be necessary for deposition and electrical contact of cathodically-active MnO_x at the metal surface so that electron transfer from the metal to the MnO_x deposit could occur.

The resulting ennoblement, observed under laboratory conditions, mimicked the pattern of ennoblement of stainless steels submerged in natural waters. However, the ennoblement produced in the laboratory study was not accompanied by the characteristic pitting corrosion of the metal, demonstrating the limitations of our current understanding of pit initiation and propagation in steels by MDB.

Slime-producing Bacteria

Microorganisms that produce copious quantities of EPS during growth in biofilms have been implicated in localized attack of stainless steels (92). Slime-forming microorganisms that have been recovered from sites of corrosion on stainless steels include *Clostridium* spp., *Flavobacterium* spp., *Bacillus* spp., *Desulfovibrio* spp.,*Desulfotomaculum* spp. and *Pseudomonas* spp.

As little as 10 ng cm^{-2} EPS has been reported to provoke the onset of MIC of stainless steel in natural seawater; cathodic protection of the stainless steel, used to prevent corrosion, actually increased the amount of EPS in the biofilm (97). However, the role of EPS in MIC of stainless steel remains obscure. It has been postulated that they are not sufficient to induce biocorrosion of stainless steel unless aided by the presence of a biocatalyst of oxygen reduction (98), which could be oxido-reductase enzymes entrapped in the biofilm (58). EPS has even been suggested to protect metal surfaces from corrosion. A bacterial consortium consisting of a thermophilic*Bacillus sp.* and *Deleya marina* produced metal-binding EPS that reduced the rate of corrosion of carbon steel by 94% (35). Such a mechanism may be responsible for the protection microorganisms afford to mild steel under certain conditions (102).

A case of corroded copper pipework in a drinking water system involved the presence of a film that stained positive with periodic acid- Schiff's reagent (PAS) and alcian blue, suggesting the presence of acidic polysaccharides (2). Scanning electron microscopy showed that copious amounts of biofilm were associated with the pitted sites (55), with the most severely corroded tubes containing the well-developed biofilm (76). In another case, chemical analysis of the adherent copper corrosion products recovered from failed copper tube suggested an interaction between the inorganic products and biologically-derived organic molecules. Copper corrosion products were located on top of or within a microbial biofilm layer in direct contact with the bare metal surface in areas where the pipe was perforated (38, 39). The biofilm contained linear and/or cross-linked acidic or non-ionic polysaccharides, oligopeptides and Nacetylated derivatives of glucose, mannose and galactose. Corrosion products rich in copper complexes of pyruvate, acetate, and histidine were identified (89). Binding of $[Cu_2Cl_2]_n^{2-}$ ions in the biofilm suggested a mechanism whereby Cl^- sequestration into the pits could promote further ionization of metallic copper (38). Microbiological evaluation of the corrosion deposits showed that while high numbers of bacteria were associated with the pits, the presence of bacteria was not always related to pitting and that the range of cultured bacterial species was quite variable (110, 111).

A correlation has been reported between pitting of copper pipe associated with a black cupric oxide surface layer and the presence of certain bacteria (*Pseudomonas paucimobilis* and *Ps. solanacearum*) and their polysaccharide (2, 21). Davidson *et al.* (28) correlated the production of acidic metabolic products by a biofilm of the bacterium *Acidovorax delafieldii* on a copper surface to an increase in copper concentration in the bulk aqueous phase (i.e., corrosion). The amount of extractable, surface-associated copper was positively correlated with both protein and carbohydrate concentrations in the biofilm. Bremer and Geesey (17) showed a correlation between acidic polysaccharide accumulation in bacterial biofilms on copper films and initiation of copper film dissolution.

Little *et al.* (68) used scanning vibrating electrode microscopy, employing a 20 μm microprobe, to demonstrate the formation of localized anodic areas on copper coupons in the presence of the marine bacterium*Oceanospirillum* and its exopolymer. Fluorescence microscopy with the Live/Dead Backlight Viability Kit®showed that the anodic areas corresponded to those with higher bacterial densities, but the sequence in which the surface changes occurred was not determined.

The relationship between pitting propensity and the properties of biofilm polymers has been investigated by Siedlarek *et al.* (100). Cyclic voltammetry showed that the artificial biofilms formed by the model polysaccharides, xanthan, alginate and agarose, displayed cation selectivity and exerted considerable influence on the corrosion reaction(s) of a copper surface in contact with an aqueous phase, particularly at the sites where solid corrosion products were precipitated (100, 113). A physicochemical model was developed to describe the pitting corrosion observed on copper piping of potable water systems. The model takes into account membrane properties and heterogeneity, and the distribution of exopolymers on the surface of the pipes (113).

Acid-producing Bacteria (APB)

Bacteria can produce copious quantities of either inorganic or organic acids as by-products of metabolism Acidophilic sulfur oxidizing bacteria (SOB), such as *Thiobacillus* spp., oxidize reduced forms of sulfur to sulfate. These microbes can cause severe corrosion damage to mining equipment. Organic acid-producing bacteria were suggested as the primary cause in a case of carbon steel corrosion in an electric power station; they were the only group of culturable microorganisms whose abundance was correlated positively with corrosion (103). Acetic, formic and lactic acids are common metabolic by-products of APB. Little *et al.* (62) showed that an aerobic, acetic acid-producing bacterium accelerated the corrosion of cathodically protected stainless steel. Protective calcium-rich deposits formed during cathodic polarization were destabilized or dissolved by artificially applied acetic acid. Little *et al.* (65) also provided examples of acids synthesized in the Krebs Cycle, common to most aerobic microorganisms, which can contribute to MIC; however, the intermediate metabolites of the Krebs cycle are generally retained within the microbial cells. A culture of *Streptococcus* released high amounts of copper from a Cu-Zn-Al-Ni odontological alloy (91), showing that lactic acid released by these bacteria can participate in corrosion reactions.

The mechanism of action of acids on corrosion of mild steel is well established in the metallurgical literature (99), but the acids produced and their concentrations are rarely monitored under MIC conditions. Acids produced by slime-producing microorganisms are concentrated at the metal surface; hence the bulk aqueous phase pH (most frequently measured by investigators) may be an entirely irrelevant parameter. Microsensors have been used to probe the pH gradients within 1mm thick microbial biofilms growing on corroded mild steel surfaces (59) pH values increased from 7.5 at the bulk fluid-biofilm interface to 9.5 at the metal surface in cathodic areas and ranged from 5 to 7 at the surface of the tubercle in anodic areas.

Slime-producing microorganisms that excrete acidic extracellular polysaccharides during biofilm formation on metal surfaces may influence corrosion. Carboxylic acid groups of matrix polysaccharides such as alginic acid, produced by the biofilm-forming bacterium *Pseudomonas aeruginosa,* have been calculated to be of the order of 6 Angstroms apart, and thus highly concentrated at the metal-biofilm interface (52). It is virtually impossible to concentrate dissolved low molecular weight acids to such a high level. These ionizable acidic groups may therefore be very important in corrosion when the pH of the biofilm is low.

Fungi

Fungi are well-known to produce organic acids, and are therefore capable of contributing to MIC. Much of the published work on biocorrosion of aluminum and its alloys has implicated fungal contaminants of jet fuel, *Hormoconis* (previously classified as *Cladosporium*) *resinae, Aspergillus spp., Penicillium spp.* and *Fusarium spp.* The fungus *H. resinae* utilizes the hydrocarbons of fuel to produce organic acids. Surfaces in contact with the aqueous phase of fuel-water mixtures and sediments are common sites of attack (95). The large quantities of organic acid by-products excreted by this fungus selectively dissolve or chelate the copper, zinc and iron at the grain boundaries of aircraft aluminum alloys, forming pits which persist under the anaerobic conditions established under the fungal mat. Growth of this and other fungi in diesel fuel storage tanks can produce large quantities of biomass (13) and this may provoke crevice attack on the metal (37). Grease-coated wire rope wound on wooden spools stored in a humid environment has been reported to be corroded by *Aspergillus niger* and *Penicillium spp.* Both fungal species are known to produce citric acid (67), which may be involved in the attack.

Iron-reducing fungi have been isolated from tubercles in a water distribution system (36), suggesting another mechanism whereby corrosion may be accelerated by this group of microorganisms.

Microbial Consortia

Microorganisms are almost never found in nature as pure species and, while laboratory studies on isolated pure cultures are essential to the understanding of MIC, the role of microbial consortia is becoming increasingly recognized.

The acids produced by APB serve as nutrients for SRB and methanogens and it has been suggested that SRB proliferate at sites of corrosion due to the activities of APB (103). Dowling *et al.* (33) compared corrosion of C1020 pipeline steel in the presence and absence of the acetogenic bacterium, *Eubacterium limosum*, and mixed SRB populations (*Desulfovibrio sp.* and *Desulfobacter spp.*). *E. limosum* alone had little effect on the corrosion rate compared to sterile controls, but when inoculated with the *Desulfovibrio sp.*, a significantly higher rate of corrosion was found. It was proposed that by-products of *E. limosum* supported *Desulfovibrio* sp. growth and sulfide production.

The interactions between microbial species are complex. Gaylarde and Johnston (42) showed that anaerobic corrosion of mild steel was enhanced in pure cultures of *Desulfovibrio vulgaris*, but reduced to below control levels by pure *Vibrio anguillarum*; in the presence of both species, corrosion rates were the highest of all. On the other hand, a second facultatively anaerobic bacterium, probably of the genus *Citrobacter*, had little effect on corrosion rates, except in triple cultures, where it apparently modified the action of the other species (Fig. 2). It was suggested that *V. anguillarum* produced a strongly-bound, protective film on the metal surface in pure cultures, but that this film incorporated SRB cells when *D. vulgaris* was present, turning it into a highly aggressive biofilm. The incorporation of the third organism into this biofilm would reduce the SRB population, thereby ameliorating its effects.

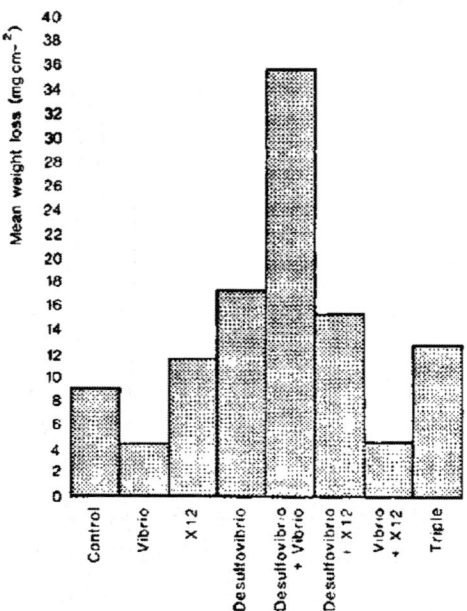

Figure 2: Weight loss of mild steel exposed to pure and mixed bacterial cultures after 3 weeks incubation in Postgate Medium B at room temperature (approx. 22. °C). X12 = presumptive *Citrobacter.*

Consortia of MDB and SRB often exist as biofilms on corroding metal surfaces. It has been proposed that oxygen consumption by MDB creates redox conditions favorable for the growth of SRB (106) and the joint action of MDB and SRB may promote the breakdown of the passive film on stainless steel (61).

A bacterial consortium was shown to be necessary for the maintenance of corrosion current of pitted 304L stainless steel in seawater under anaerobic conditions (3). SRB were present on the cathode, leading to high charge transfer resistance, while the consortium on the anode decreased charge transfer resistance. These results were stated to support the involvement of cathodic depolarization in the anaerobic biocorrosion of stainless steel.

A number of microorganisms isolated from corroding copper pipework in Auckland, New Zealand, attached to and grew on copper surfaces in a simulated potable water medium (114). The four most numerous culturable bacterial species were identified by 16s rRNA gene sequence analysis as *Sphingomonas capsulata* (European Bioinformatics Institutes (EMBL) Nucleotide sequence database # AJ223450), *Staphylococcus warneri* (EMBL #AJ223451), *Erythrobacter longus* (EMBL # AJ223452) and *Methylobacterium sp.* (EMBL #AJ223453) A yeast, identified as a *Candida sp.*, was also recovered from the copper surface. Biofilms containing these isolates were shown to promote release of copper corrosion by-products in subsequent laboratory reactor experiments (114).

Techniques for the Study of Biocorrosion

The forms of corrosion which can be promoted by the interaction of microorganisms with metals are numerous, including general pitting, crevice attack, stress corrosion cracking, enhancement of corrosion-fatigue, intergranular stress cracking and hydrogen embrittlement and cracking. Most cases of MIC are associated with localized attack (Fig. 3). The complexity of MIC reactions means that a broad range of techniques must be employed to relate the corrosion processes to the microbial activities at surfaces.

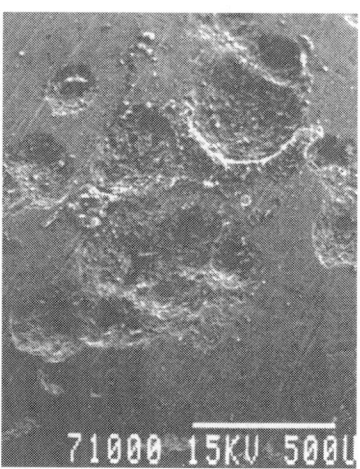

Figure 3: SEM micrograph of mild steel surface, showing localized attack, following exposure to mixed population of *Pseudomonas* spp and sulfate-reducing bacteria.

Qualitative and Semi-quantitative Evaluation of MIC

The contribution of microorganisms to corrosion has been assessed using a variety of optical and electron microscopy techniques. Recently, environmental scanning electron microscopy (ESEM), atomic force microscopy (AFM; Fig. 4) and confocal laser scanning microscopy (CLSM) have been employed to study biofilms and biocorrosion phenomena (10). Microscope techniques provide information about the morphology of microbial cells and colonies, their distribution on the surface, the presence of EPS (Figs. 1 and 4) and the nature of corrosion products (crystalline or amorphous; Fig. 5a and b). They can also reveal the type of attack (e.g. pitting or uniform corrosion) by visualizing changes in microstructure and surface features after removal of the biofilm and corrosion products (Fig. 6). CLSM and AFM allow the examination of hydrated biofilms and yield clean, three-dimensional images of living biofilms in real time. CLSM has shown that 75 to 95% of the volume of bacterial biofilms is occupied by the matrix, and cells may be concentrated in only 5-25% of the lower or upper layers (24). ESEM studies of biocorrosion and protective coatings have also been reported (54, 112). However, the detection of microorganisms, in itself, should not be the sole basis on which their involvement in the corrosion process is implicated. To confirm MIC, specific activities of the microbes at the site where corrosion is occurring should be demonstrated. Microscopic and culture techniques alone rarely provide such evidence.

Figure 4: Atomic force microscopy image of a single bacterial cell and its associated EPS on a surface of AISI 316 stainless steel.

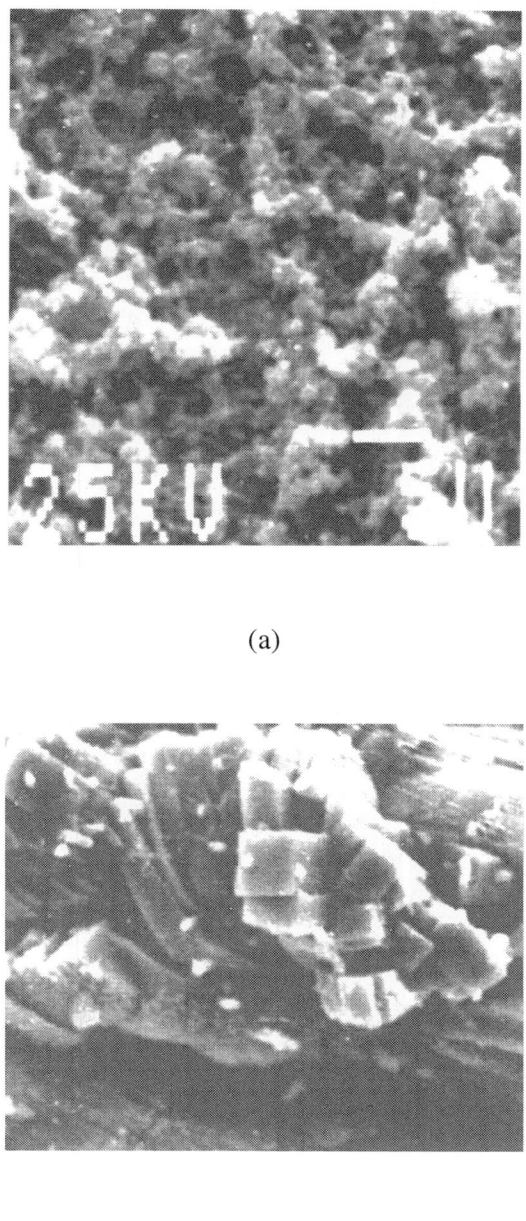

(a)

(b)

Figure 5: SEM micrograph of amorphous (a) and crystalline (b) biocorrosion products on a mild steel surface.

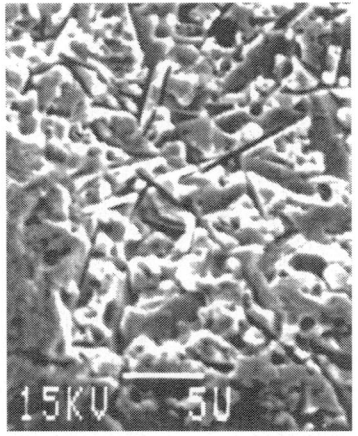

(a)

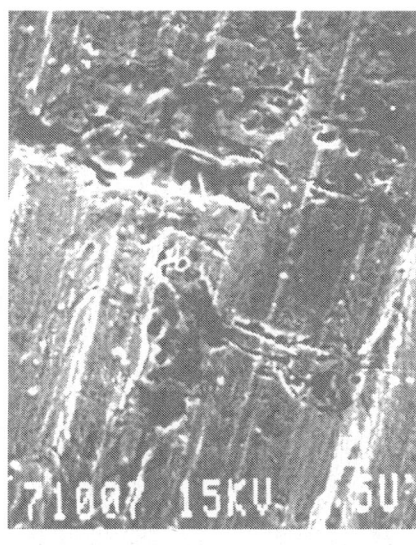

(b)

Figure 6: SEM image of a mild steel surface after the removal of bacterial biofilm, revealing changes in surface characteristics.

Chemical spectroscopy at surfaces offers information on the nature of the accumulated corrosion products, which can be specifically associated with microbial activities. Spatially resolved surface chemistry obtained by spectroscopy must be related to the spatially resolved microbiology at the same location. Surface chemical analysis provides information on the chemical composition of the corrosion products and microbiological deposits, and thus gives the opportunity to gain insight into the electrochemical reactions involved in the corrosion process. X-ray diffraction (XRD) and energy dispersive X-ray analysis (EDAX) have been widely used to obtain elemental information on corrosion products on metal surfaces (82). Auger electron spectroscopy (AES) allows mapping of corrosion products across a metal surface that has experienced localized attack. It has been used to investigate biocorrosion in condenser tubes (22). X-ray photoelectron spectroscopy (XPS) can resolve the oxidation state of the elements present, facilitating prediction of corrosion product chemistry and, to some extent, chemistry of the associated microbial biofilm (90). It has also been used to determine the influence of a biofilm on the structure of the passive layer formed on AIS 316 stainless steel (11). AES and XPS are suitable only for evaluating the composition of thin scaling deposits, but laser Raman spectroscopy (LRS), coupled with optical microscopy, can be used to analyze thicker (above 1 μm) deposits (101) and offers an interesting technique for future biocorrosion studies. Mössbauer spectroscopy can be applied to iron-containing compounds. It has been used to detect «green rust 2» among corrosion products of steel exposed to marine sediments containing SRB (88) and subsequent, controlled laboratory studies showed that this corrosion product was exclusively associated with SRB-induced corrosion (83).

Machado *et al.* (80) used XRD, Mössbauer spectrophotometry and EDAX to show that the surface film formed on mild steel in the presence of a consortium of *H. resinae* and SRB was mainly composed of magnetite (Fe_3O_4), maghemite ($g-Fe_2O_3$), goethite ($a-FeOOH$) and lepidocrocite ($g-FeOOH$). Under the experimental conditions used, this film was protective.

The presence of mackinawite and greigite among corrosion products of iron is generally evidence that SRB participated in the corrosion reaction (51, 77, and 79). Under alternating reducing and oxidizing conditions, the partially oxidized iron oxide magnetite is often produced, along with lepidocrocite and goethite (51). These mineral signatures of MIC have been detected, using XRD and EDAX, as corrosion products on many oil and gas pipeline systems (51). Amorphous iron sulfide is also often detected by EDAX at pipeline corrosion sites. Little is known about its subsequent crystallization, although biomineralization around SRB colonies or within biofilms may be a key process.

Characteristic copper sulfides, chalcocite (Cu_2S), covellite (CuS_{1-x}) and djurleite ($Cu_{31}S_{16}$) are formed during corrosion of copper and its alloys in the presence of SRB (77, 78). The formation of thick, non-adherent layers of chalcocite or the formation of hexagonal chalcocite is indicative of SRB-induced corrosion of copper and copper alloys.

Quantitative Assessment of MIC

Corrosion rates are commonly determined by electrochemical methods, such as potentiodynamic polarization, zero-resistance ammetry, and electrochemical impedance spectroscopy (EIS) and electrochemical noise (ECN), in addition to classical weight loss measurements. A detailed review of these techniques is given by Mansfield and Little (81).

Microsensors, which are largely electrochemically-based, offer the resolution that is needed for studying the localized corrosion processes induced by microorganisms. They have been applied to characterize the chemical gradients within biofilms on corroding metal surfaces. Microsensors were employed to show depletion of oxygen within tubercles formed on corroding mild steel surfaces (59) and at anodic areas of the surface covered by a 1mm-thick biofilm. This spatially resolved surface chemical approach enabled these investigators to demonstrate the existence of differential oxygen concentration cells and their role as the driving force for the corrosion reaction.

Attenuated total reflectance Fourier transform infrared spectroscopy (ATR/FTIR) was used to quantify the rate of corrosion of a thin metal film deposited as an internal reflectance element, which is exposed to either flowing or stagnant aqueous media. The method is based on the observation that water absorbance in the infrared increases as the thin film decreases in thickness as a result of corrosion. Changes in film thickness corresponding to a few atomic layers can be detected and the measurements can be obtained non-destructively in real time. Quantitative changes in water absorbance are expressed as a corrosion rate. ATR/FTIR has been used to demonstrate the participation of a microbial biofilm in the localized attack of copper films and the relation between onset of corrosion and the production of polysaccharide during biofilm formation (17). It has also been used to demonstrate the influence of the exopolysaccharide produced by the marine bacterium*Pseudoalteromonas* (*Pseudomonas*) *atlantica* on the corrosion of copper (53).

It is indisputable that both qualitative and quantitative approaches are necessary to investigate the role of microorganisms in corrosion processes. Increasingly sophisticated techniques are being employed to study corrosion, microbial activities in biofilms and the types of microorganisms present. The information from the use of molecular gene probes, demonstrating that the majority of microorganisms in the natural environment are unculturable, means that our understanding of MIC is extremely limited. This multi-disciplinary subject, with its important practical applications, is certain to be an area of intense research activity in the future.

ACKNOWLEDGEMENTS

We wish to thank CNPq for a grant (Pesquisador Visitante) to I B Beech.

REFERENCES

1.   Abdollahi, H.; Wimpenny, J.W.T. Effects of oxygen on the growth of *Desulfovibrio desulfuricans*. *J. Gen. Microbiol.*, 136: 1025-1030, 1990.

2.   Angell, P.; Campbell, H.S.; Chamberlain, A.H.L. International Copper Association (ICA), Project N° 405, Interim Report, 1990.

3.   Angell, P.; Luo, J.-S.; White, D.C. Microbially sustained pitting corrosion of 304 stainless steel in anaerobic seawater. *Corr. Sci.*, 37: 1085-1096, 1995.

4.   Bak, F.; Cypionka, H.A. novel type of energy metabolism involving fermentation of inorganic sulfur compounds. *Nature*, 326: 891-892, 1987.

5.   Barton, L.L. (ed.) *Sulfate-reducing bacteria*. Plenum Press, New York, 1984.

6.   Beech, I. B.; Cheung, C.W.S.; Chan, C.S.P.; Hill, M.A.; Franco, R.; Lino A.R. Study of parameters implicated in the biodeterioration of mild steel in the presence of different species of sulphate-reducing bacteria. *Internat. Biodet. Biodeg*, 34: 289-303, 1994.

7.   Beech, I.B.; Cheung, C.W.S. Interactions of exopolymers produced by sulphate-reducing bacteria with metal ions. *Internat. Biodet. Biodeg*, 35: 59-72, 1995.

8.   Beech, I.B.; Zinkevich, V.; Tapper, R.; Gubner, R.; Avci, R. The interaction of exopolymers produced by marine sulphate-reducing bacteria with iron, in: W. Sand (ed.), *Biodeterioration and Biodegradation*, 10th International Biodeterioration and Biodegradation Symposium, Hamburg, Dechema Monographs vol.133, Frankfurt, 1996, pp. 333-338.

9.   Beech, I.B.; Zinkevich, V.; Tapper, R.; Gubner, R The direct involvement of extracellular compounds from a marine sulphate-reducing bacterium in deterioration of steel *Geomicrobiology J.*, 15: 119-132, 1998.

10.   Beech, I.B.; Tapper, R.; Gubner, R. Microscopy methods for studying biofilms. In: *Biofilms: recent advances in their study and control*. L.V. Evans (ed.), Harwood Academic Publ. 1999. In press.

11.   Beech, I.B.; Gubner, R.; Zinkevich, V.; Hanjangsit, L.; Avci, R The effect of *Pseudomonas* NCIMB biofilm on the formation of passive layer on AISI 316 stainless steel *Corrosion 99*, NACE, Houston Tx., Paper N° 185, 1999a.

12.   Beech, I.B., Zinkevich, V., Tapper, R. and Avci, R. Study of the interaction of exopolymers produced by sulphate-reducing bacteria with iron using X-ray Photoelectron Spectroscopy and Time-of-Flight Secondary Ionisation Mass Spectrometry. *J. Microbiol. Meths* 36, 3-10, 1999.

13.   Bento, F.M.; Gaylarde, C. Microbial contamination of stored diesel oil in Brazil. *Rev. Microbiol.* 27:192-196, 1996.

14.   Bibb, M. Bacterial corrosion in the South African power industry. In S.C. Dexter (ed.) *Biologically Induced Corrosion*, National Association of Corrosion Engineers, Houston, TX, 1986, pp. 96-101.

15. Booth, G.H. Sulphur bacteria in relation to corrosion. *J. Appl. Bacteriol.*, 27: 174-181: 1964.

16. Borenstein, S.W. *Microbially influenced corrosion handbook.* Woodhead, Cambridge, UK, 1994.

17. Bremer, P.J.; Geesey, G.G. Laboratory-based model of microbially induced corrosion of copper. *Appl. Environ. Microbiol.*, 57: 19561962, 1991.

18. Brennenstuhl, A.M.; Doherty, P.E The economic impact of microbiologically influenced corrosion at Ontario Hydro's nuclear plants. In: N.J. Dowling, M.W. Mittelman and J.C. Danko (eds.), *Microbiologically Influenced Corrosion and Biodeterioration*, University of Tennessee, Knoxville, TN, 1990. pp. 7/5-7/10.

19. Bryant R.D.; Jansen, W.; Bovin, J.; Laishley, E.J.; Costerton, J.W. Effect of hydrogenase and mixed sulfate reducing bacterial populations on the corrosion of steel. *Appl. Environ. Microbiol.*, 57: 2804-2809, 1991.

20. Bryant, R.D.; Kloeke, F.V.O.; Laishley, E.J. Regulation of the periplasmic [Fe] hydrogenase by ferrous iron in *Desulfovibrio vulgaris* Hildenborough. *Appl. Environ. Microbiol.*, 59: 491-495, 1993.

21. Chamberlain, A.H.L.; Angell, P.; Campbell, H.S. Staining procedures for characterising biofilms in corrosion investigations. *Br. Corros. J.*, 23: 197-199, 1988.

22. Chen, J-R.; Chyou, S.D.; Lew, S.-I.; Huang, C.-J.; Fang, C.-S.; Tse, W.-S. Investigation of the biological corrosion of condenser tubes by scanning Auger microprobe techniques *Appl Surf Sci.*, 33/34: 212-219, 1988.

23. Cheung, C.W.S.; Beech, I.B. The use of biocides to control sulphate-reducing bacteria in biofilms on mild steel surfaces *Biofouling*, 9: 231-249, 1996.

24. Costerton, J.W. Structure of Biofilms. In: *Biofouling and Biocorrosion in Industrial Water Systems*. Lewis, Boca Raton, Fla., 1994, p. 1.

25. Cragnolino, G.; Tuovinen, O.H The role of sulfate-reducing and sulfur-oxidising bacteria in the localized corrosion of iron-based alloys, a review. *Internat. Biodet.*, 20: 9-18, 1984.

26. Crolet, J.-L From biology and corrosion to biocorrosion, *Oceanol. Acta*, 15: 87-94, 1992.

27. Daumas, S.; Massiani, Y.; J. Crousier. Microbiological battery induced bysulphate-reducing bacteria. *Corr. Sci.*, 28: 1041-1050, 1988.

28. Davidson, D.; Beheshti, B.; Mittelman, M.W. Effects of *Arthrobacter* sp., *Acidovorax delafieldii* and *Bacillus megatherium* colonization on copper solvency in a laboratory reactor. *Biofoul.*, 9: 279-292, 1996.

29. Davies, D.G.; Parsek, M.R.; Pearson, J.P.; Iglewski, B.H.; Costerton, J.W.; Greenberg, E.P The involvement of cell-to-cell signals in the development of a bacterial biofilm. *Science*, 280, 295-298. 1998.

30. Dickinson,W.; Lewandowski, Z. Manganese biofouling and corrosion behaviour of stainless steel. *Biofouling*, 10: 79-93, 1996.

31. Dickinson, W.F.; Caccavo, J.R.; Olesen, B.; Lewandowski, Z. Ennoblement of stainless steel by the manganese-depositing bacterium *Leptothriix discophora. Appl. Environm. Microbiol.*, 63: 2502-2506, 1997.

32. Dilling, W.; Cypionka, H. Aerobic respiration in sulfate-reducing bacteria *FEMS Micriobiol. Lett.*, 71: 123-128, 1990.

33. Dowling, N.J.E.; Brooks, S.A.; Phelps, T.J.; White, D.C. Effects of selection and fate of substrates supplied to anaerobic bacteria involved in the corrosion of pipe-line steel. *J. Ind. Microbiol.*, 10: 207-215, 1992.

34. Dowling, N.J.E.; Mittelman, M.W.; White, D.C. The role of consortia in micobially influenced corrosion. In: J.G. Zeikus, (ed.), *Mixed Cultures in Biotechnology*, McGraw Hill, New York, 1991, pp. 341-372.

35. Edyvean, R.G.J.; Benson, J.; Thomas, C.J.; Beech, I.B.; Videla, H. Biological influences on hydrogen effects in steel in seawater. *Mat Perform.* 37: 40-44, 1998.

36. Emde, K.M.E.; Smith, D.W.; Facey, R. Initial investigation of microbially influenced corrosion (MIC) in a low temperature water distribution system. *Water-Research*, 26: 169-175, 1992.

37. Englert, G.E.; Müller, I.L. Corrosion behavior of martensitic steel in diesel oil. In: *LABS 3*, eds. C.C. Gaylarde, T.C.P. Barbosa and N.H. Gabilan, The British Phycological Society, UK, 1998, Paper n° 24.

38. Fischer, W.R.; Hänssel, I.; Paradies, H.H. First results of microbial induced corrosion on copper pipes. In:*Microbial Corrosion-1*, C.A.C. Sequeira and A.K. Tiller (eds), Elsevier Applied Sciences, London, 1988, pp. 300-327.

39. Fischer, W.R.; Paradies, H.H.; Wagner, D.; Hän_el, I. Copper deterioration in a water distribution system of a county hospital in Germany caused by microbially induced corrosion. 1. Description of the problem. *Werkst. Korros.*, 43: 56-62. 1992.

40. Flemming, H.-C. Biofouling and microbiologically influenced corrosion (MIC) an economical and technical overview. In: E. Heitz, W. Sand and H.-C. Flemming (eds.) *Microbial Deterioration of Materials*, Springer, Heidelberg, 1996, pp. 514.

41. Gaylarde, C.C.; Johnston, J.M The effect of *Vibrio anguillarum* on the anaerobic corrosion of mild steel by *Desulfovibrio vulgaris*. *Internat. Biodeter. Bull.*, 18: 111-116, 1982.

42. Gaylarde, C.C.; Johnston, J.M. Anaerobic metal corrosion in cultures of bacteria from estuarine sediments. In: S.C. Dexter (ed), *Biologically Induced Corrosion*, NACE8, Houston, TX, 1986, pp. 137-143.

43. Gounot, A.M. Microbial oxidation and reduction of manganese: Consequences in groundwater and applications *FEMS Microbiol. Rev.*, 14: 339-350, 1994.

44. Gubner, R. J.; Breakspear, S.; Beech, I. B The effect of iron and dissolved oxygen level on the biocorrosion of piling grade carbon steel. *LABS 3*, eds. C.C. Gaylarde, T.C.P. Barbosa and N.H. Gabilan, The British Phycological Society, UK, 1998, Paper n° 41.

45. Hamilton, W.A. Sulphate reducing bacteria and anaerobic corrosion. *Ann. Rev. Microbiol.*, 39: 195-217, 1985.

46. Hamilton, W.A. Metabolic interaction and environmental microniches: implications for the modeling of biofilm process. In: G.G. Geesey, Z. Lewandowsky and H.-C. Flemming (eds.) *Biofouling and Biocorrosion in Industrial Water Systems*, 2nd edition, CRC Press Inc., Boca Raton, FL, 1994, pp. 27-36.

47.   Hamilton, W.A. Sulphate-reducing bacteria: physiology determines their environmental impact.*Geomicrobiol. J.*, 15: 19-28, 1998.

48.   Hardy, J.A.; Hamilton, W.A. The oxygen tolerance of sulfate-reducing bacteria isolated from the North Sea waters. *Curr. Microbiol.*, 6: 259-262, 1981.

49.   Iverson, I.P.; Olson, G.J. Anaerobic corrosion by sulfate-reducing bacteria due to highly reactive volatile phosphorus compound. In: *Microbial Corrosion*, Metals Society, London, 1983, pp. 46-53.

50.   Jack, R.F.; Ringelberg, D.B.; White, D.C. Differential corrosion rates of carbin steel by combinations of*Bacillus* sp., *Hafnia alvei*, and *Desulfovibrio gigas* established by phospholipid analysis of electrode biofilm.*Corr. Sci.*, 33: 1843-1853, 1992.

51.   Jack, T.R.; Wilmott, M.J.; Sutherby, R.L. Indicator minerals formed during external corrosion of line pipe.*Mat Perform*. 34: 19-22, 1995.

52.   Jang, L.K.; Quintero, E.; Gordon, G.; Rohricht, M.; Geesey, G.G. The osmotic coefficients of the sodium form of some polymers of biological origin. *Biopolymers*, 28: 1485-1489, 1989.

53.   Jolley, J.G.; Geesey, G.G.; Hankins, M.R.; Wright, R.B.; Wichlacz, P.L *In situ*, real time FT IR/CIR/ATR study of the biocorrosion of copper by gum arabic, alginic acid, bacterial culture supernatant and *Pseudomonas atlantica* exopolymer. *Appl. Spectrosc.*, 43:1062-1067, 1989.

54.   Jones-Meehan, J.; Walch, M.; Little, B.J.; Ray, R.I.; Mansfeld, F.B. *Biofouling and Biocorrosion in Industrial Water Systems*. Lewis, Boca Raton, Fla. 1994, p. 107.

55.   Keevil, C.W.; Walker, J.T.; McEvoy, J.; Colbourne, J.S. Detection of biofilms associated with pitting corrosion of copper pipework in Scottish hospitals. In: C.C. Gaylarde and L.H.G. Morton, (eds)., *Biocorrosion*.Biodeterioration Society Occasional Publication N° 5, Biodeterioration Society, Lancashire, U.K., 1989. pp. 99-117.

56.   King, R.A.; Wakerley, D.S. Corrosion of mild steel by ferrous sulphide. *Br. Corr. J.*, 8: 41-45, 1973.

57.   Kobrin, G. Corrosion by microbiological organisms in natural waters. *Mat Perform*. 15: 38-43, 1976.

58. Lai, M.E.; Scotto, V.; Bergel, A. Analytical characterization of natural marine biofilms *10th Internat. Congress on Marine Corrosion and Fouling, Melbourne, Australia, Feb., 1999* In press.

59. Lee, W.; deBeer, D. Oxygen and pH microprofiles above corroding mild steel covered with a biofilm.*Biofouling*, 8: 273-280, 1995.

60. Lee, W.; Lewandowski, Z.; Nielsen, P.H.; Hamilton, W.A. Role of sulfate-reducing bacteria in corrosion of mild steel: A review. *Biofouling*, 8: 165-194, 1995.

61. Lewandowski, Z.; Dickinson, W.; Lee, W. Electrochemical interactions of biofilms with metal surfaces. *Wat. Sci. Tech.*, 36: 295-302, 1997.

62. Little, B.; Wagner, P.; Duquette, D. Microbiologically-induced increase in corrosion current density of stainless steel under cathodic protection. *Corrosion*, 44: 270-274, 1988a.

63. Little, B.; Wagner, P.; Jacobus, J The impact of sulfate-reducing bacteria on welded copper-nickel seawater piping systems. *Mat. Perform*. 27: 57-61, 1988b.

64. Little, B.; Wagner, P.; Ray, R.; McNeil, M Microbiologically influenced corrosion in copper and nickel seawater piping systems. *Mar. Technol. Soc. J.*, 24: 10-17, 1990.

65. Little, B.; Wagner, P.; Mansfeld, F An overview of microbiologically influenced corrosion. *Electrochim. Acta*, 37: 2185-2194, 1992.

66. Little, B.; Wagner P. Indicators for sulfate-reducing bacterium microbiologially influenced corrosion. In: G.G. Geesey, Z. Lewandowski and H.C. Flemming (eds.) *Biofouling and Biocorrosion in Industrial Water Systems*, 2nd ed, CRC Press Inc., Boca Raton, FL, 1994, pp. 213-230.

67. Little, B.; Ray, R.; Hart, K.; Wagner, P. Fungal-induced corrosion of wire rope. *Mater Perform*. 34: 55-58, 1995.

68. Little, B.J.; Wagner, P.A.; Angell, P.; White, D. The role of bacterial exopolymer in marine copper corrosion.*LABS2 - Biodegradation and Biodeterioration in Latin America*, eds. C.C. Gaylarde, E.L.S. de Sa and P.M. Gaylarde, UNEP/UNESCO/ICRO-FEPAGRO/ UFRGS, Brazil, 1996, pp. 98-100.

69. Little, B. J.; Wagner, P.; Mansfeld, F. Microbiological testing In *Microbiologically Influenced Corrosion. Corrosion testing made easy*, vol.5, NACE International, Houston, TX, 1997a, pp. 29-52.

70.   Little, B.J.; Wagner, P.; Hart, K.; Ray, R.; Lavoie, D.; Nealson, K.; Aguilar, C. The role of metal-reducing bacteria in microbiologically influenced corrosion. In: *Proc. NACE Corrosion `97*, NACE International, Houston, TX, Paper N° 215, 1997b.

71.   Little, B.J.; Wagner, P.A.; Lewandowski, Z The relationship between biomineralization and microbiologically influenced corrosion. *LABS 3*, eds. C.C. Gaylarde, T.C.P. Barbosa and N.H. Gabilan, The British Phycological Society, UK, 1998, Paper n° 50.

72.   Little, B.J.; Ray, R.I.; Jones-Meehan, J. Interpreting spatial relationships between marine bacteria and localized corrosion on steels *10th Internat Congress on Marine Corrosion and Fouling, Melbourne, Australia, Feb., 1999* in press.

73.   Little, J.; Edyvena, R.G.J The influence of marine fouling on hydrogen permeation through steels. *Proc. NSF CONICET workshop Biocorrosion and Biofouling*, eds. H.A. Videla, Z. Lewandowski and R.W. Lutey, Buckman Intl., Memphis, 1992.

74.   Lovley, D.R.; Philips, E.J.P. Novel processes for anaerobic sulfate production from elemental sulfur by sulfate-reducing bacteria. *Appl. Environ. Microbiol.*, 60: 2394-2399, 1994.

75.   Lutey, R.W. Identification and detection of microbiologically influenced corrosion. In: H.A. Videla, Z. Lewandowski; R. Lutey (eds), *Proc. NSF-CONICET Workshop Biocorrosion and Biofouling, Metal/Microbe Interactions*, November 92, Mar del Plata, Argentina, Buckman Laboratories International, Inc., Memphis, TN, 1992, pp. 146-158.

76.   McEvoy, J.; Colbourne, J.S. Glasgow hospital survey pitting corrosion of copper tube. *Report to the International Copper Research Association, New York*, 1988.

77.   McNeil, M.B.; Jones, J.M.; Little, B.J. Mineralogical fingerprinting for corrosion processes induced by sulfate reducing bacteria Paper N° 580, *Proc. NACE Corrosion `91*, NACE, Houston, TX, 1991a.

78.   McNeil, M.B.; Jones, J.M.; Little, B.J. Production of sulfide minerals by sulfate-reducing bacteria during microbiologically influenced corrosion of copper. *Corrosion*, 47: 674-677, 1991b.

79.   McNeil, M.B.; Little, B.J. Technical note: Mackinawite formation during microbial corrosion. *Corrosion*, 46: 599-600, 1990.

80. Machado, P.F.L.; Gaylarde, C.C.; Müller, I.L. Microbial influenced corrosion of mild steel ASTM A283 in a marine diesel-aqueous system. *LABS 3*, eds. C.C. Gaylarde, T.C.P. Barbosa and N.H. Gabilan, The British Phycological Society, UK, 1998, Paper n° 54.

81. Mansfield, F.; Little, B.J A critical review of the application of electrochemical techniques to the study of MIC. In: N.J. Dowling, M.W. Mittleman and J.C. Danko, (eds.), *Microbially Influenced Corrosion and Biodeterioration*, University of Tennessee, Knoxville, TN, 1991, pp. 5/33-5/40.

82. Marquis, F.D.S. Strategy of macro and microanalysis in microbial corrosion. In: C.A.C. Sequeira and A.K. Tiller, (eds.), *Microbial Corrosion1*, Elsevier Applied Science, London and New York, 1988, pp. 125-151.

83. Morton, L.H.G.; Greenaway, D.L.A.; Gaylarde, C.C.; Surman, S.B. Consideration of some implications of the resistance of biofilms to biocides. *Internat. Biodeter. Biodeg.*, 41: 247-259, 1998.

84. Moulin, J. M.; Marsh, E.; Chao, V.; Karius, R.; Beech, I. B.; Gubner, R.; Raharinaivo, A. *Prevention of Accelerated Low-Water Corrosion on Steel Piling Structures due to Microbiologically Influenced Corrosion Mechanisms*; ECSC final draft report, ECSC agreement 7210KB/503, KB/825, KB/130, KB/826, KB/337, 1998.

85. Myers, C.; Nealson, K.H. Bacterial manganese reduction and growth with manganese oxide as the sole electron acceptor. *Science*, 240: 1319-1321, 1988.

86. Obuekwe, C.O.; Westlake, D.W.S.; Plambeck, J.A.; Cook, F.D. Corrosion of mild steel in cultures of ferric iron reducing bacterium isolated from crude oil I. polarization characteristics. *Corrosion*, 37 (8): 461-467, 1981.

87. Olowe, A.; Benbouzid-Rollet, N.D.; Génin, J-M.R.; Prieur, D.; Confente, M; Resiak, B. La présance simultanée de rouille verte 2 et de bactéries sulfato- réductrices en corrosion perforante de palplanches en zone portuaire, *C.R. Académie des Sciences Paris*, t 314, Série II, 1992, pp. 1157-1163.

88. Olowe, A.; R.Génin; J-M.; Guezennec, J. Mössbauer effect study of microbially induced corrosion of steel by sulphate reducing bacteria in marine sediments: role of green rust 2. In: N.J. Dowling, M.W. Mittleman and J.C. Danko, (eds.), *Microbially Influenced Corrosion and Biodeterioration*, University of

Tennessee, Knoxville, TN, 1991, pp. 5/65-5/72.

89. Paradies, H.H.; Fischer, W.R.; Haenssel, I.; Wagner, D. Characterisation of metal biofilm interactions by extended absorption fine structure spectroscopy. In: C.A.C. Sequeira and A.K. Tiller (eds), *Microbial Corrosion-2*European Federation of Corrosion, Publication N° 8, The Institute of Materials, 1992, pp 168-188.

90. Pendyala, J.; Avci, R.; Geesey, G.G.; Stoodley, P.; Hamilton, M.; Harkin, G. Chemical effects of biofilm colonization on 304 stainless steel. *J. Vac. Sci. Technol.*, A 14: 1755-1760, 1996.

91. Pizzolitto, E.L.; Pizzolitto, A.C.; Bernardi, A.C.A.; Lochagin, N.; Shuhama, T.; Ito, I.Y.; Guastaldi, A.C. Biocorrosão: estudo *in vitro* da aderência microbiana, nas ligas metálicas de cobre de aplicação odontológica.*2nd NACE Latin American Region Corrosion Congress* September, 1996, Paper LA 96108.

92. Pope, D.H.; Duquette, D.J.; Johannes, A.H.; Wayner, P.C. Microbially influenced corrosion of industrial alloys. *Mat Perform.* 23: 14-18, 1984.

93. Postgate, J. R *the Sulphate Reducing Bacteria* 2nd edn Cambridge University Press, England 1984.

94. Roden, E.E.; Lovley, D.R. Dissimilatory Fe (III) reduction by marine microorganism *Desulfuromonas acetooxidans*. *Appl. Environ. Microbiol.*, 59: 734-742, 1993.

95. Salvarezza, R.C.; Videla, H.A. Electrochemical behaviour of aluminium in *Cladosporium resinae* cultures.*Biodeterioration 6*, S. Barry et al. (Eds.), London, Elsevier, 1986, 1986, pp. 212-217.

96. Schaschl, E. Elemental sulfur as a corrodent in dearated neutral aqueous environments. *Mat. Perf.*, 19: 9-11, 1980.

97. Scotto, V. Microbial and biochemical factors affecting the corrosion behaviour of stainless steels in seawater. *A working party report on marine corrosion of stainless steels: chlorination and microbial effects.* N° 10, Institute of Materials, London, 1993, pp. 21-33.

98. Scotto, V.; Alabiso, A.; Beggiato, M.; Marcenaro, G.; Guezenneck, J. Possible chemical and microbiological factors influencing stainless steel MIC in natural seawater. In: C. Christiansen, (ed.), *Proceedings of the 5thEuropean Congress on Biotechnology, Copenhagen, Denmark*, vol 2, 1990, pp. 866-871.

99. Shreir, L.L The microbiology of corrosion. In: *Corrosion*, Vol. 1, J. Wiley, NewYork, 1963. pp. 2.52-2.64.

100. Siedlarek, H.; Wagner, D.; Fischer, W.R.; Paradies, H.H. Microbiologically influenced corrosion of copper: the ionic transport properties of bioploymers. *Corr. Sci.*, 36: 17511763, 1994.

101. Singh Raman, R.K.; Gleeson, B.; Young, D.J. Laser Raman spectroscopy: a technique for rapid characterisation of oxide scale layers *Mats. Sci. Technol.*, 14, 373-376, 1998.

102. Soracco, R.J.; Berger, L.R.; Berger, J.A.; Mayack, L.A.; Pope, D.H.; Wilde, E.W. Microbiologically mediated reduction in the pitting of mild steel overlaid with plywood *Proc. NACE Corrosion `84*, NACE, Houston, TX, Paper N° 98, 1984.

103. Soracco, R.J.; Pope, D.H.; Eggers, J.M.; Effinger, T.N microbiologically influenced corrosion investigations in electric power generating stations *Proc NACE Corrosion `88*, NACE, Houston, TX, Paper N° 83, 1988.

104. Tatnall, R. Fundamentals of bacterial induced corrosion. *Mat. Perform.* 20: 32-38, 1981.

105. Thomas, C.J.; Edyvean, R.G.J.; Brook, R. Biologically enhanced fatigue. *Biofouling* 1: 65-77, 1988.

106. Videla, H.A.; Characklis, W.G. Biofouling and microbially influenced corrosion. *Internat. Biodegrad. Biodeter.*, 29: 195-207, 1992.

107. Videla, H.A.; Mele, M.F.L.; Swords, C.; Edyvean, R.G.J.; Beech, I.B. Comparative study of the corrosion product films formed in biotic and abiotic media. *Corrosion 99*, NACE, Houston, Tx., Paper N° 163, 1999.

108. Videla, H.A.; Swords, C.L.; de Mele, M.; Edyvean, R.G.; Watkins, P.; Beech, I.B The role of iron in SRB-influenced corrosion of mild steel. *Corrosion 98*, NACE, Houston, Tx., Paper N° 289, 1998.

109. Von Wolzogen Kuhr, C.A.H.; van der Vlugt, L.S; De grafiteering van Gietijzer als electrobiochemisch Proces in anaerobe Grunden. *Water (den Haag)* 18: 147-151, 1934.

110. Wagner, D.; Fischer, W.R.; Paradies, H.H. Copper deterioration in a water distribution system of a county hospital in Germany caused by microbially induced corrosion. 2. Simulation of the

corrosion process in two test rigs installed in this hospital. *Werkst. Korros.*, 43: 496-500, 1992a.

111. Wagner, D.; Fischer, W.R.; Paradies, H.H. Test methods on microbial induced corrosion in different loops In *Proceedings of the 12th Scandinavian Corrosion Congress and Eurocorr*, Dipoli, Finland, 1992b, pp. 651-665.

112. Wagner, P.; Little, B.; Ray, R.; Jones-Meehan, J. Investigation of microbiologically influenced corrosion using environmental scanning electron microscopy *Proc NACE Corrosion `92*, NACE, Houston, TX, Paper N° 185, 1992.

113. Wagner, D.; Chamberlain, A.H.L.; Fischer, W.R.; Wardell, J.N.; Sequeira, C.A.C. microbiologically influenced corrosion of copper in potable water installations - a European project review. *Mat Corr.*, 48; 311-321, 1997.

114. Webster, B.J.; Wells, D.B.; Bremer, P.J The influence of potable water biofilms on copper corrosion *Proc NACE Corrosion `96*, NACE, Houston, TX, Paper N° 294, 1996.

115. Werthén, M.; Hermansson, M.; Elwing, H. In-vitro studies of signal-receptor interactions in bacterial cell density dependent AHL-signaling *10th Internat. Congress on Marine Corrosion and Fouling, Melbourne, Australia, Feb. 1999* In press.

116. White, D.C.; Nivens, D.E.; Nichols, P.D.; Mikell, A.T.; Kerger, B.D.; Henson, J.M.; Geesey, G.G.; Clarke, K.C. Role of aerobic bacteria and their extracellular polymers in the facilitation of corrosion. In: S.C. Dexter (ed.), *biologically induced corrosion*, NACE8, NACE, Houston, TX, 1986, pp. 233-243.

117. Widdel, F. Microbiology and ecology of sulfate and sulfur-reducing bacteria. In: A.J.B. Zehnder (ed.), *Biology of Anaerobic Microorganisms*, Wiley-Liss, John Wiley and Sons, Inc., New York, 1988, pp. 469-586.

Electrochemical Tests to Evaluate the Stability of the Anodic Films on Dental Implants

C. E. B. Marino[1] and L. H. Mascaro[2]

[1]Laboratório de Materiais Bicompatíveis, Departamento de Engenharia Mecânica, Universidade Federal do Paraná, Jardim das Américas, CP 19081, 81531-990 Curitiba, PR, Brazil

[2]Laboratório Interdisciplinar de Cerâmica, Departamento de Química, Universidade Federal de São Carlos, Rod. Washungton Luiz, km 235, CP 676, 13565-905 São Carlos, SP, Brazil

ABSTRACT

The stability of anodic films potentiodynamically grown on titanium, titanium-grade 2, and Ti6Al4V alloy was studied in a simulated

physiological electrolyte, up to 8.0 V, and at room temperature to determine the corrosion resistance levels of dental implants. In PBS (phosphate buffer saline) solution, thin titanium oxide films protect the surface of the Ti6%Al4%V alloy up to 6.0 V, pure Ti up to 8.0 V, and Ti-grade 2 up to 1.5 V. At more positive potentials, localized corrosion starts to occur possibly due to the alloy elements (Ti6Al4V-V and Al) and variable levels of interstitials (Ti-grade 2: C, N, and Fe, mainly). When the biomaterials were submitted to open-circuit conditions, in artificial saliva, the worst corrosion resistance was observed in dental implant (Ti-grade 2), according to the open-circuit potential values and reconstruction rate analysis of these oxide films. The XPS spectra revealed TiO_2 oxide as the main phase in the barrier oxide film coating the dental implant.

INTRODUCTION

The use of titanium as a material for surgical purposes began around the 1960s. For this purpose, this material must not cause any adverse biological reaction in the body, and it must also be stable and retain its functional properties. Due to its attractive mechanical and chemical characteristics and its property of "biofixation" with the periprosthetic tissue, titanium is established as one of the major materials used for manufacturing orthopedic implants [1]. The excellent corrosion resistance of titanium and its alloys results from the formation of very stable, continuous, highly adherent, and protective oxide films on metal surfaces. Because titanium itself is highly reactive and has extremely high affinity for oxygen, these beneficial surface oxide films form spontaneously and instantly when fresh metal surfaces are exposed to air. The nature, composition, and thickness of the protective surface oxides on titanium alloys depend on environment conditions. Titanium is normally covered with a thin oxide film, which largely determines the surface properties of an implant [1]. The high corrosion resistance of titanium is due to the formation of these protective oxide films (thickness 1 to 4 nm) [2], but body fluids contain chloride ions that can induce a breakdown of passive films on prostheses [3,4]. Corrosion and surface film dissolution are two mechanisms for introducing additional ions in the human body. Extensive release of metal ions from prosthesis can result in adverse biological reactions and can lead to mechanical failure of the device [5].

The thin oxide film, naturally formed on a titanium substrate, is responsible for the excellent biocompatibility of titanium implants. This fact occurs because of the low level of electronic conductivity, a thermodynamically stable state in the physiological media, besides high corrosion resistance [6]. Previous studies have described a variety of properties of anodic oxide films, for example, surface morphology, crystal structures, and corrosion resistance [7]. One way to change the surface characteristics of implants is varying the properties of the oxide films always present on Ti surfaces. Oxidized implants change the properties of the titanium implant and play an important role during the dynamic buildup of the osteointegration process [1, 8]. Healy and Ducheyne [9] already investigated the superficial changes, stoichiometry, and absorbed surface species on titanium oxide during exposure to model physiological environments. As a result of these studies, it was observed that the oxide incorporates both Ca and P elements from the extra cellular fluid and that it increases in thickness as function of implantation time. Kuphasuk et al. [10] studied the electrochemical behavior of Ti, Ti6Al4V, NiTi, and three other titanium alloys at 37°C in Ringer's solution. Electron diffraction patterns indicated that all titanium alloys were covered mainly with rutile-type oxide after corrosion tests, and the Ti and Ti5Al2.5Fe were the most resistant to corrosion. The study of the corrosion resistance in chloride media of Ti-based dental implants was investigated by Arslan et al. [11]. By the Tafel slopes and EIS Nyquist spectra, the galvanic corrosion was evaluated, and the Ti6Al4V/CoCr and Ti6Al4V/CrNi systems were to be similar corrosion behavior.

One of the problems associated with corrosion study of dental alloys is that the knowledge of the electrochemical environment of oral cavity is limited to simulating this media. The electrochemical properties of the oxide film and its long-term stability in biological environments play a decisive role for the biocompatibility of titanium implants [12–14].

A method of expanding the corrosion resistance of titanium into reducing environments is increasing the surface oxide film thickness by anodizing or thermal oxidation [1]. In the present work, the growth of the oxide and its stability in Ti, Ti-grade 2, and Ti6Al4V alloy were studied in physiological solutions at room temperature (~25°C) by electrochemical techniques. The titanium oxide characterization was made through X-ray photoelectron spectroscopy (XPS) and microstructural aspects by metallographic techniques.

EXPERIMENTAL

A conventional 250 mL three-electrode electrochemical cell was used. The working electrodes were Ti6%Al4% V (6% m/m and 4% m/m of Al and V, resp.) alloy rod (Titanium Industries-USA), pure Ti (Aldrich, 99.99%), and Ti-grade 2 (dental screw [15]) giving an area of ~1.26 cm^2, ~0.28 cm^2 and 0.062 cm^2, respectively, and were exposed to the electrolytes. Prior to the oxide growth, the electrodes were mechanically polished with silicon carbide paper of grade 600 and rinsed with distilled water. A platinum electrode was used as counterelectrode, and all potentials were measured against a saturated calomel electrode (SCE). The electrolytes were PBS solution (phosphate buffer saline-pH 7; 8.77 g·L^{-1} NaCl, 3.58 g·L^{-1} Na$_2$HPO$_4$, 1.36 g·L^{-1} KH$_2$PO$_4$) [16] for the growth of the oxides and artificial saliva (pH 7; 0.4 g·L^{-1} NaCl, 0.4 g·L^{-1} KCl, 0.795 g·L^{-1} CaCl$_2$·H$_2$O, 0.69 g·L^{-1}NaH$_2$PO$_4$·H$_2$O, 0.005 g·L^{-1} Na$_2$S·9H$_2$O, 1.0 g·L^{-1} urea) [17] for the study of spontaneous dissolution process, at room temperature. All electrochemical measurements were carried out with a Microquimica-MQPG-01 electrochemical system. The initial (E$_i$) and final (E$_f$) potentials for the potentiodynamic growth at 50 mV·s^{-1}of the oxides were E$_i$ =-1.0 V and E$_f$ 1.0-8.0 V (versus SCE) range. The electrodes were polished and the oxides potentiodynamically grown up to E$_f$, where the potential was held until a steady current was reached. Therefore, the circuit was open and the variation of potential with time (dissolution process) was analyzed for the Ti and its alloys covered with oxide electrodes, in artificial saliva solution. So, the stability of the anodic oxide in artificial saliva was studied. Finally, the anodic scan was repeated, in PBS solution, in order to obtain the potentiodynamic profile for the regrowth of any oxide that eventually had been dissolved [2]. The nature of the anodic oxides was investigated by X-ray photoelectron spectroscopy (XPS) using a Kratos Analytical XSAM HS Spectrometer having Mg Kα (hn =1253.6eV), X-ray source with power given by the emission of 15 mA at a voltage of 15 kV. The high-resolution spectra were obtained with analyzer energy of 20 eV. The binding energies were referred to the carbon 1s line, set at 284.8 eV. A least-square routine was used for the fitting of the peaks.

RESULTS AND DISCUSSION

Anodic Growth and Stability of the Oxides

Cyclic voltammetry was used to characterize the pure Ti system, varying E_f between 1.0 V and 8.0 V. This potential range was chosen because the corrosion process starts at potentials higher than 6.0 V, for Ti6Al4V alloy, in PBS solution [18, 19], and higher than 1.5 V, for Ti-grade 2, in the same solution, as will be discussed. The voltammogram shown in Figure 1 indicates that the Ti-grade 2 has a similar anodic transformation when compared with pure Ti [2]. This fact is possible due to the growth of a uniform oxide film, mainly TiO_2. However, the potential sweep in the positive direction shows an anodic current starting with –0.55 V for the titanium electrode [2] and –0.15 V for Ti-grade 2 electrode. The delay in starting potential oxide growth, in the case of dental implant (Ti-grade 2), could be explained by the presence of impurities and/or variable levels of interstitials into the biomaterial that causes differences in their stability. These hypotheses will be analyzed by microstructural and by chemical composition tests. With an increase in the potential, the current density remains almost constant indicating a process of thickening of the anodic surface film. After the growth and aging of the oxide films up to different potentials (E_f), the circuit was open and the potential decaying was measured, to analyze the spontaneous dissolution of these different oxide films. From this, the immersion time in artificial saliva was varied. When the open-circuit potential (OCP) became stationary, a new potential sweep, from – 1.0 to 1.5 V, was carried out to regrow any film that might have been dissolved. In acid solutions, Blackwood et al. [20] proposed that the dissolution reaction of open-circuit breakdown of passive film on Ti might be written as

$$TiO_2 + H_2O + H^+ \longrightarrow Ti(OH)_3$$

$$(1)$$

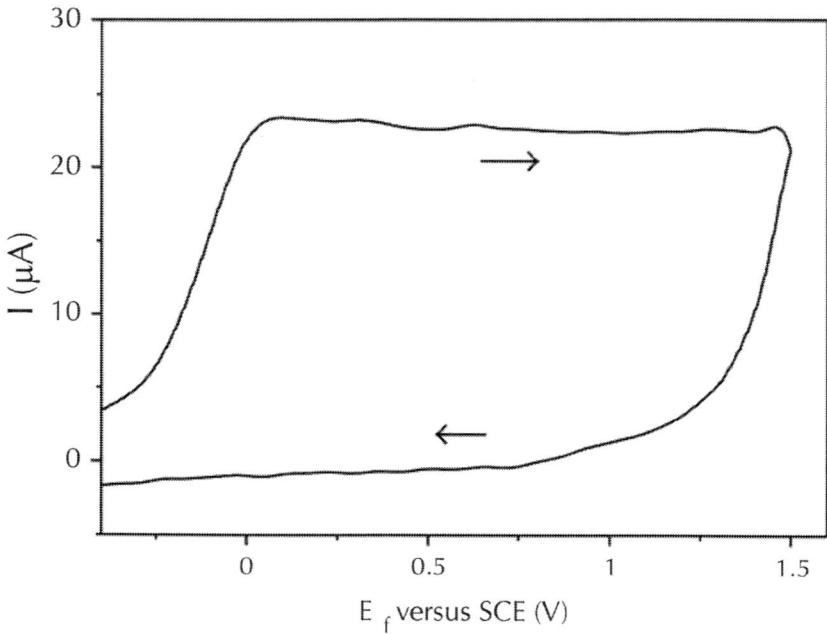

Figure 1: Cyclic voltammogram for Ti-grade 2 electrode in PBS solution at 50 mV·s^{-1} and 25°C.

Following the OCP evolution, it is possible to understand the general interactions and the spontaneous dissolution of oxide grown up to 1.5 V that takes place at the biomaterials surface. The open-circuit potential (E_{oc}) of the pure Ti, Ti-grade 2, and Ti6Al4V alloy in artificial saliva is shown in Figure 2. In artificial saliva, the E_{oc} of all biomaterials, was stabilized after a week at values between –0.63 V and –0.24 V versus SCE. The worst corrosion resistance was observed in the dental implant (Ti-grade 2) according to lower E_{oc} and higher reconstruction rate (see below).

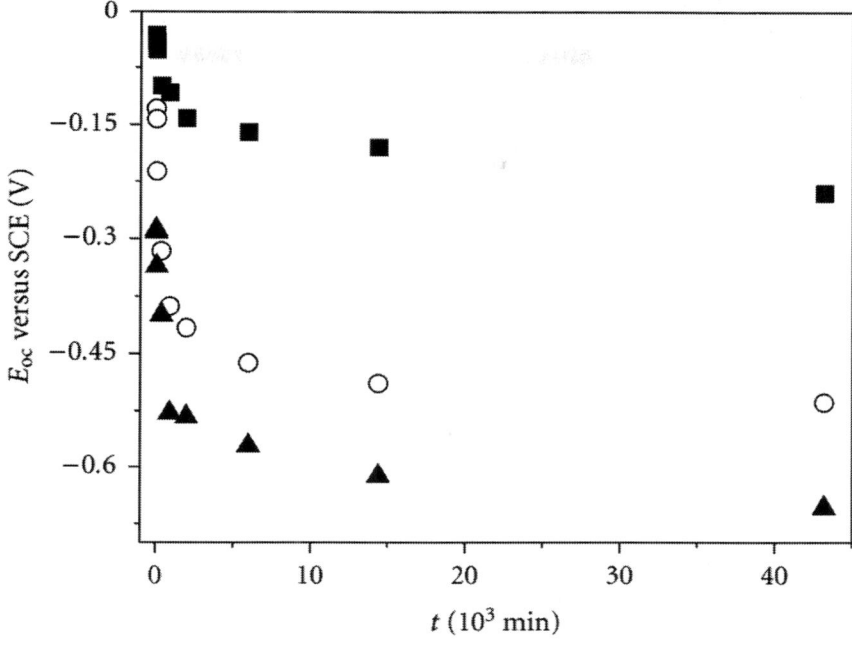

E_f: 1.5 V
- ■ Pure-Ti
- ○ Ti6Al4V
- ▲ Ti-grade 2

Figure 2: Open-circuit potentials as a function of immersion time, in artificial saliva, for Ti pure, Ti6Al4V, and Ti-grade 2 covered with oxide films potentio-dynamically grown up to 1.5 V, in PBS solution.

The anodic charges related to the formation (Q_f) and reconstruction Q_{rec} of the anodic oxide allowed for an evaluation of the magnitude of these processes, in PBS solution. These charges were obtained considering the areas under the anodic sweeps before and after OCP measurements in artificial saliva, which are integrated between −1.0 and 1.5 V. It means that the amount of current consumed while the electrochemical dissolution of the oxide film during E_{oc} was integrated. Figure 3 shows a typical valve metal behavior for the Ti-grade 2, and its anodic oxide grows according to the high-field model [2].

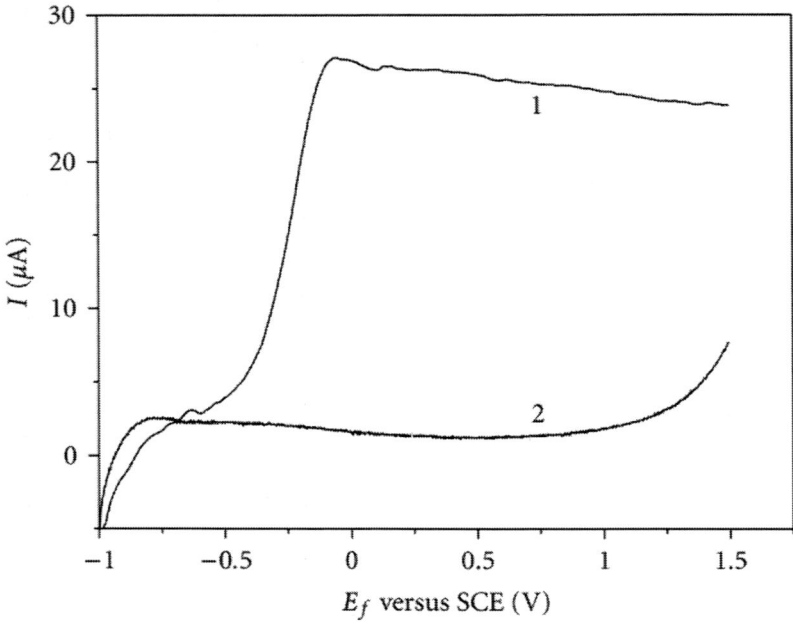

Figure 3: Voltammetric profiles for the growth (curve 1) and regrowth (curve 2) of the oxide film grown on Ti-grade 2 up to E_f 1.5 V, in PBS solution, at 50 mV·s⁻¹ and 25°C.

The oxide film reconstruction rate (RR, expressed as a percentage) is defined as being equal to the ratio between the Q_{rec} and the Q_f [2], and these anodic charges were obtained by integration of the corresponding I versus E curves. Figure 4 shows the dependence of the reconstruction rate (RR%) on immersion time in artificial saliva. The RR values of oxides increase with immersion time and stabilize in 15% Ti-grade 2, 13% Ti6Al4V alloy, and 9% pure Ti, after around a week, in artificial saliva. So, the dissolution process of these oxide films is more effective in the dental implant (Ti-grade 2) in this physiological media as previously observed by potential sweep. Species from the electrolytic solution can be incorporated into the oxide matrix [20] leading to changes in its mechanical properties. The oxides have a natural level of internal stress, but the presence of incorporated foreign species affects directly the Pilling-Bedworth ratio (ratio between the oxide molar volume and the metal molar volume), inducing alterations in the internal stress that may cause the rupture of the protecting oxide film [21].

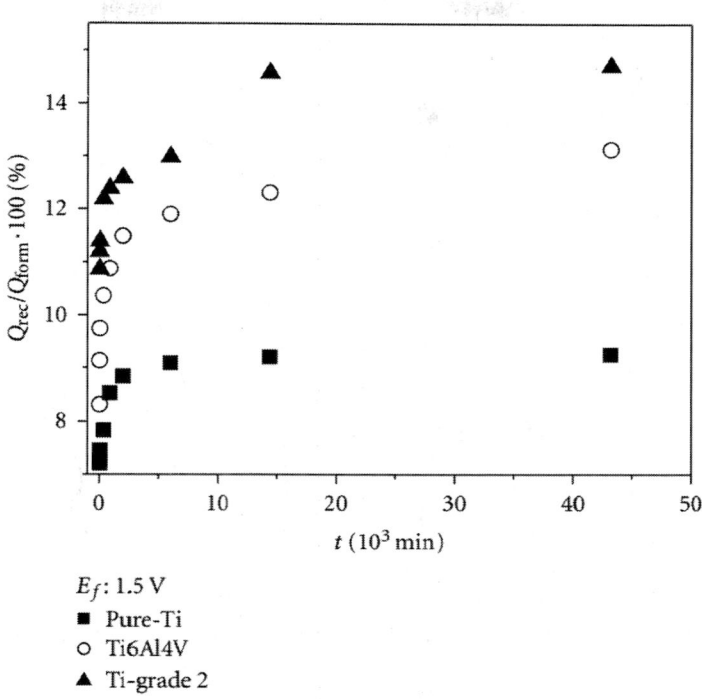

E_f: 1.5 V
■ Pure-Ti
○ Ti6Al4V
▲ Ti-grade 2

Figure 4: Oxide film reconstruction rate (RR%) as a function of immersion time, in saliva artificial, for Ti pure, Ti6Al4V alloy, and Ti-grade 2 electrodes covered with oxide films potentiodynamically grown up to 1.5 V, in PBS solution.

In agreement with what was previously stated, the impedance data previously described [11, 19, 22] show the influence of the Cl⁻ presence on the passive oxide stability. By completing this research, we already presented the characterization of these anodic films by electrochemical impedance [19]. The impedance response indicated the existence of a thin oxide film on dental implant that suffers a dissolution process during different immersion times. This fact was confirmed by resistive and capacitive behaviors, as the corrosion resistance of the Ti-grade 2 is less than pure Ti. The resistance of the passive film on Ti-grade 2 decreased when the exposure time is increased, due to the oxide breakdown followed by a dissolution process. The capacitive behavior indicated the presence of a uniform and thin oxide on the implant surface. The influence of chloride media on the stability of anodic films on titanium-based materials by voltammetric studies has been presented by our research members [18].

An average value of 2.5 nm V^{-1} was obtained for the anodization rate in pure Ti oxides [2]. In the case of Ti-grade 2, the anodization rate is 2.1 nm V^{-1} indicating a lower protect layer to a fixed final potential. This result is according to the magnitude of the reconstruction rate obtained to the dental implant in artificial saliva (see Figure 4). This worst corrosion resistance could be related to the presence of contaminants and impurities and/or to the microstructural variations that occur after thermal and conformational treatments. Therefore, the dental implant becomes more vulnerable to the corrosion processes [23].

In the corrosion-dissolution process, the presence of ions released from the implant could affect the tissue-bone material interactions that cause damage in the osteointegration process resulting from the looseness of implant. Marino and Mascaro [19] have found that the spontaneous dissolution rate of TiO_2 in sulfuric acid solutions, at various pHs, is first order with respect to proton concentration. Thus, we proposed that the oxide hydrolysis might be written as reaction (1). Therefore, a very slow chemical dissolution could be associated with the release of $Ti(OH)_3^+$ ions and other unstable elements in oxide matrix.

MICROSTRUCTURAL ANALYSIS

The electrochemical results suggest that the corrosion resistance decreases with increasing levels of impurities. Therefore, the microstructural analysis could show the natural tendency of biomaterial corrosion resistance.

Information about the chemical composition and structural aspects of these biomaterials is essential to the knowledge of its behavior and stability in aggressive media, such as body fluids [24, 25]. The microstructures of the Ti, Ti-grade 2 and Ti6Al4V alloy were analyzed after mirror polishing and etching their surfaces with a Kroll solution ($6 : 1 : 1000$ v/v/v $HNO_3 : HF : H_2O$) for approximately 15 s. The adherent oxide film on the surface of Ti and Ti alloys requires strong etchings to minimize the staining problems that might obscure the microstructure.

Titanium has hexagonal close-packed atomic lattice structure (α titanium-phase) which is stable up to 882°C, and it transforms into a face-centered cubic structure (β titanium phase) above this temperature. Pure titanium is a soft and ductile metal, but the impurities or addition of alloying elements makes it harder and less ductile. In the Ti6Al4V alloy, the aluminum stabilizes the α phase and the vanadium stabilizes β phase. These alloying elements are known to increase the mechanical properties of titanium [1]. The addition of the elements aluminum and vanadium refined the grain size of the alloy [18]. The pure titanium microstructure shows a satisfactory homogeneity, indicating good corrosion resistance [20] (see Figure 5(a)).

(a)

(b)

Figure 5: Photomicrographs of the Ti (a) and the dental implant, Ti-grade 2 (b), surfaces etched with the Kroll reagent (200x).

Figure 5(b) shows the dental implant (Ti-grade 2) microstructure, in which the superficial deposition of the attack product is not uniform. This fact occurs because of the different expositions of the crystallographic planes, and each grain has its own crystalline orientation in relation to the attack plane. Comparing the pure Ti and Ti-grade 2 microstructures, it is possible to observe some important alterations. The Ti-grade 2 microstructure presents nonuniform grains. The elongated grains indicate that the material was deformed, without recrystallization, and sustain some strain hardening (tensioned regions) (see Figure 5). Also, notice that dark regions (more intensive attack) are in major proportions than clear regions, which confirms the partially strain-hardened material. These strain-hardened areas are anodic regions in corrosion process.

The difference in Ti purity levels and its fabrication characterizes the kinds of Ti grade 1, 2, 3, and 4 and their mechanical and corrosion resistance properties [1]. Thus, the microstructural analysis sustained that the lowest corrosion resistance of Ti-grade 2 is due to the levels of interstitials and other impurities, a fact that damages the osteointegration process [3].

CHARACTERIZATION OF OXIDE FILM

The characterization of the passive oxide film coating Ti-grade 2 implant was done by X-ray photoelectron spectroscopy (XPS). The potential value was chosen in order to have passive films grown up to 1.5 V (~40 Å thick, according to previous results [2]), in PBS solution. Figure 6 presents the XPS spectrum for Ti 2p electron to the dental implant covered with titanium oxide film obtained in phosphate buffer solution. The spectrum shows the doublet referent to Ti 2p that consists at the peaks: Ti $2p_{1/2}$ and Ti $2p_{3/2}$. These subindices indicate the level of the element in spectroscopic analysis. The doublet was determined by the Ti $2p_{3/2}$ peak with binding energy around 459.0 eV, which corresponds to the Ti^{4+} signal [26]. This indicates that TiO$_2$ is the main oxide present in the passive film, consistent with results reported previously using different approaches [18, 27].

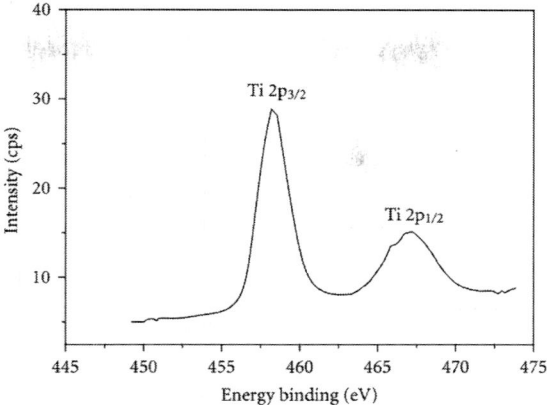

Figure 6: Ti 2p XPS spectrum for the oxide film potentiodynamically grown on the dental implant, Ti-grade 2, in PBS solution.

When the surface analyzed was Ti6Al4V/oxide, the effects of alloying elements in passive films were examined by X-ray photoelectron spectroscopy (Figure 7). The peaks indicate that TiO_2 (Ti^{4+}) is the major surface component for both Ti—grade 2 and Ti6Al4V alloy samples. The peak positions for TiO_2 were consistent with binding energy shifts expected for titanium in combination with oxygen in the form of TiO_2 [21, 27].

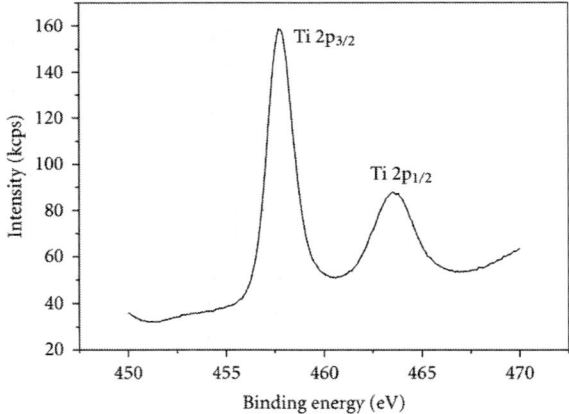

(a)

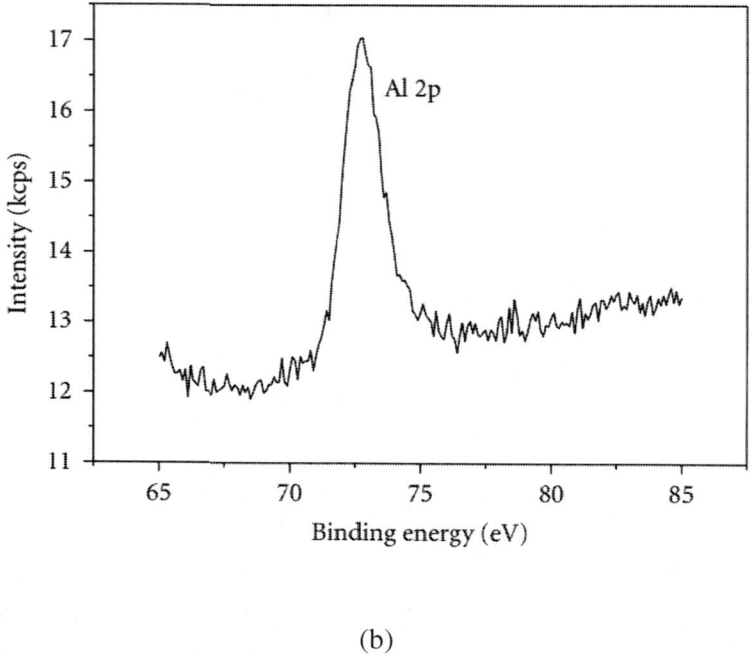

(b)

Figure 7: (a) Ti 2p and (b) Al 2p XPS spectra for the oxide film potentiody-namically grown on the Ti6Al4V alloy, in PBS solution.

The high-resolution Al 2p spectrum results for the Ti6Al4V alloy sample are shown in Figure 7(b). The presence of elemental aluminum metal can be verified (Al^0) at binding energy at 72.5 eV. Some authors [27,28] reported the measured binding energies of surface Al and V to be consistent with oxides such as Al_2O_3 and V_2O_5 or even Al and V existing as ions at interstitial or substitutional sites in the Ti oxide matrix. In the present work, the thin oxide film on the Ti6Al4V alloy could be a "solid solution" of titanium oxide (TiO_2) that contains dissolved aluminum and vanadium. In addition, by the chemical environment analysis of Al, which indicated the absence of O atoms around it, and considering the values of ionic radius of these alloy elements, it was possible to suppose that both Al and V would be present as ions at interstitial sites in the Ti oxide matrix. The vanadium alloy element was not detected by XPS analysis but this element could be present in the oxide with concentrations lower than 0.1 at.% which represents the detection limit of the XPS technique.

Besides, the total XPS spectrum revealed Fe contaminant present in the Ti-grade 2 oxide matrix. The binding energy of Fe 2p peak is around 715.0 eV, and its atomic percentage is 1.8% which could affect its properties. For pure Ti and Ti6Al4V alloy, the absence of Fe was observed [28]. Finally, it was described that high levels of interstitials in dental implants affected the corrosion resistance and the osteointegration process [3, 5].

CONCLUSIONS

In this work, it was shown that thin titanium oxide films grown potentiodynamically protect the Ti-grade 2 surfaces up to 1.5 V in artificial saliva, at room temperature. At more positive potentials, corrosion processes start to occur. This can be explained due to the presence of variable levels of interstitials (impurities) in Ti-grade 2 screws that could decrease their corrosion resistance. The results were confirmed by metallographic and XPS data. The XPS spectra revealed the TiO_2 as the most important phase in the passive film, besides the presence of Fe traces that modify the dental implant properties. Therefore, Ti-grade 2 screws showed an increase of heterogeneity regarding its chemical composition and microstructural aspects when compared with pure Ti. These changes cause susceptibility to corrosion processes, in body fluids. So, the presence of oxide films coating the implant surface is a good pretreatment against corrosion processes, and it also increases the durability of these implants.

ACKNOWLEDGMENTS

The authors are grateful to the Conselho Nacional para o Desenvolvimento Científico e Tecnológico (CNPq) for the scholarship and to the Fundação Araucária for the financial support.

REFERENCES

1.   K. J. Bundy, "Corrosion and other electrochemical aspects of biomaterials," Critical Reviews in Biomedical Engineering, vol. 22, no. 3-4, pp. 139–251, 1994.

2.   C. E. B. Marino, E. M. Oliviera, R. C. Rocha-Filho, and S. R. Biaggio, "On the stability of thin-anodic-oxide films of titanium in acid phosphoric media," Corrosion Science, vol. 43, no. 8, pp. 1465–1476, 2001.

3.   H. L. Myshin and J. P. Wiens, "Factors affecting soft tissue around dental implants: a review of the literature," The Journal of Prosthetic Dentistry, vol. 94, no. 5, pp. 440–444, 2005. · ·

4.   S. Virtanen, I. Milošev, E. Gomez-Barrena, R. Trebše, J. Salo, and Y. T. Konttinen, "Special modes of corrosion under physiological and simulated physiological conditions," Acta Biomaterialia, vol. 4, no. 3, pp. 468–476, 2008. · ·

5.   C.-H. Chung, et al., "Electrochemical behavior of dental implant system before and after clinical use," Transactions of Nonferrous Metals Society of China, vol. 19, pp. 846–851, 2009.

6.   Y. T. Sul, C. B. Johansson, S. Petronis et al., "Characteristics of the surface oxides on turned and electrochemically oxidized pure titanium implants up to dielectric breakdown: the oxide thickness, micropore configurations, surface roughness, crystal structure and chemical composition," Biomaterials, vol. 23, no. 2, pp. 491–501, 2002. · ·

7.   M. Metikos and R. Babic, "Some aspects in designing passive alloys with an enhanced corrosion resistance," Corrosion Science, vol. 51, no. 1, pp. 70–75, 2009. · ·

8.   K. Lee, H.-C. Choe, B.-H. Kim, and Y.-M. Ko, "The biocompatibility of HA thin films deposition on anodized titanium alloys," Surface and Coatins Technology, vol. 205, supplement 1, pp. S267–S270, 2010.

9.   K. E. Healy and P. Ducheyne, "Hydration and preferential molecular adsorption on titanium in vitro," Biomaterials, vol. 13, no. 8, pp. 553–561, 1992. · ·

10.   C. Kuphasuk, Y. Oshida, C. J. Andres, S. T. Hovijitra, M. T. Barco, and D. T. Brown, "Electrochemical corrosion of titanium and titanium-based alloys," The Journal of Prosthetic Dentistry, vol. 85, no. 2, pp. 195–202, 2001. · ·

11.   H. Arslan, H. Çelikkan, N. Örnek, O. Ozan, A. E. Ersoy, and M. L. Aksu, "Galvanic corrosion of titanium-based dental implant materials," Journal of Applied Electrochemistry, vol. 38, no. 6, pp. 853–859, 2008. · ·

12. M. Songür, H. Çelikkan, F. Gökmeşe, S. A. Şimşek, N. Ş. Altun, and M. L. Aksu, "Electrochemical corrosion properties of metal alloys used in orthopaedic implants," Journal of Applied Electrochemistry, vol. 39, no. 8, pp. 1259–1265, 2009. · ·

13. C. Fleck and D. Eifler, "Corrosion, fatigue and corrosion fatigue behavior of metal implant materials, especially titanium alloys," International Journal of Fatigue, vol. 32, no. 6, pp. 929–935, 2010.

14. D. Upadhyay, M. A. Panchal, R. S. Dubey, and V. K. Srivastava, "Corrosion of alloys used in dentistry: a review," Materials Science and Engineering A, vol. 432, no. 1-2, pp. 1–11, 2006. · ·

15. ASTM F67-95, "Standard Specification for unalloyed titanium for surgical implants applications".

16. J. Y. Gal, Y. Foret, and M. A. Yadzi, "About a synthetic saliva for in vitro studies," Talanta, vol. 53, no. 6, pp. 1103–1115, 2001. · ·

17. T. Fusayama, T. Katayori, and S. Nomoro, "Corrosion of gold and amalgam placed in contact with each other," Journal of Dental Research, vol. 42, pp. 1183–1197, 1963.

18. C. E. B. Marino, N. Bocchi, R. C. Rocha-Filho, and S. R. Biaggio, "Voltammetric stability of anodic films on the Ti6Al4V alloy in chloride medium," Electrochimica Acta, vol. 51, no. 28, pp. 6580–6583, 2006. · ·

19. C. E. B. Marino and L. H. Mascaro, "EIS characterization of a Ti-dental implant in artificial saliva media: dissolution process of the oxide barrier," Journal of Electroanalytical Chemistry, vol. 568, no. 1-2, pp. 115–120, 2004. · ·

20. D. J. Blackwood, L. M. Peter, and D. E. Williams, "Stability and open circuit breakdown of the passive oxide film on titanium," Electrochimica Acta, vol. 33, no. 8, pp. 1143–1149, 1988.

21. C. E. B. Marino, P. A. P. Nascente, N. Bocchi, R. C. Rocha-Filho, and S. R. Biaggio, "XPS characterization of anodic titanium oxide films grown in phosphate buffer solutions," Thin Solid Films, vol. 468, no. 1-2, pp. 109–112, 2004. · ·

22. V. A. Alves, R. Q. Reis, I. C. B. Santos et al., "In situ impedance spectroscopy study of the electrochemical corrosion of Ti and Ti-6Al-4V in simulated body fluid at 25∘C and 37∘C," Corrosion Science, vol. 51, no. 10, pp. 2473–2482, 2009. · ·

23. C. V. Vidal and A. I. Muñoz, "Electrochemical characterisation of biomedical alloys for surgical implants in simulated body fluids," Corrosion Science, vol. 50, no. 7, pp. 1954–1961, 2008.

24. I. Milosev, M. Metikos, and H. H. Strehblow, "Passive film on orthopaedic TiAlV alloy formed in physiological solution investigated by X-ray photoelectron spectroscopy," Biomaterials, vol. 21, no. 20, pp. 2103–2113, 2000. · ·

25. N. Zaveri, M. Mahapatra, A. Deceuster, Y. Peng, L. Li, and A. Zhou, "Corrosion resistance of pulsed laser-treated Ti-6Al-4V implant in simulated biofluids," Electrochimica Acta, vol. 53, no. 15, pp. 5022–5032, 2008. · ·

26. J. F. Moulder, P. E. Stickle, P. E. Sobol, and K. D. Bomben, Handbook of X-Ray Photoelectron Spectroscopy, Perkin-Elmer Corporation, Waltham, Mass, USA, 1992.

27. B.-S. Kang, Y. T. Sul, SE. J. Oh, H. J. Lee, and T. Albrektsson, "XPS, AES and SEM analysis of recent dental implants," Acta Biomaterialia, vol. 5, no. 6, pp. 2222–2229, 2009. · ·

28. K. Bordji, J. Y. Jozeau, D. Mainard, et al., "Cytocompatibility of Ti6Al4V and Ti5Al2.5Fe alloys according to three surface treatments, using human fibroblasts and osteoblasts," Biomaterials, vol. 17, no. 9, pp. 929–940, 1996.

Chapter 5

Effect of Secondary Phase Precipitation on the Corrosion Behavior of Duplex Stainless Steels

Kai Wang Chan and Sie Chin Tjong

Department of Physics and Materials Science, City University of Hong Kong, Tat Chee Avenue, Kowloon, Hong Kong, China

ABSTRACT

Duplex stainless steels (DSSs) with austenitic and ferritic phases have been increasingly used for many industrial applications due to their good mechanical properties and corrosion resistance in acidic, caustic and marine environments. However, DSSs are susceptible to intergranular, pitting and stress corrosion in corrosive environments due to the formation of secondary phases. Such phases are induced in DSSs during the fabrication, improper heat treatment, welding process and prolonged exposure to high temperatures during their service lives. These include the precipitation of sigma and chi phases at 700–900 °C and spinodal decomposition of ferritic grains into Cr-rich and Cr-poor phases at 350–550 °C, respectively. This article gives the state-of the-art review on the microstructural evolution of secondary phase formation and their effects on the corrosion behavior of DSSs.

INTRODUCTION

Stainless steels are an important class of engineering alloys that have found widespread applications from domestic home appliances to structural components in aerospace industries. In particular, duplex stainless steels (DSSs) with austenitic (fcc) and ferritic (bcc) grains possess beneficial combinations of these two phases [1,2,3,4,5,6]. DSSs exhibit greater toughness and better weldability than ferritics. Compared with austenitic grades, DSSs have higher resistance to pitting and stress corrosion cracking. Accordingly, they are widely used in various chemical, petrochemical, food, power, transportation, pulp and paper industries as well as oil refineries.

The high corrosion resistance of DSSs derives from their high Cr content in combination with substantial additions of Mo, Ni and N. Chromium contributes to the corrosion resistance of stainless steels by forming protective Cr-oxide/hydoxide in the passive film [7]. The presence of Mo within the passive film of DSSs and synergistic effect between the oxides/hydroxides of Cr and Mo improve the film stability against pitting corrosion [8]. As recognized, Cr, Mo and Si alloying elements stabilize ferritic phase, while Ni and N are the -phase stabilizers [9]. N is an interstitial solid solution strengthener that increases the strength of DSSs [10]. It also increases the pitting resistance of DSSs and concentrates mainly at the metal-passive film interface. Manganese stabilizes the austenite, but it is not effective as Ni in stabilizing the -phase. Therefore, N is added with Mn simultaneously to DSSs to balance a decrease in the Ni content [11,12,13,14,15]. Cu addition is beneficial for enhancing the resistance of DSSs in non-oxidizing solutions. Its content is limited to ~2.5% since higher content reduces hot ductility [2]. The ferrite/austenite ratio in DSSs must be close to 50:50 by adding appropriate alloying elements for achieving desired microstructures and mechanical properties. A wide range of DSSs with different alloying elements has been developed and commercialized. The most widely used DSS is the standard 2205 grade. To further improve corrosion resistance, DSSs with higher Cr content such as 25% Cr grade (UNS S32550 and S32950), superduplex (UNS S32750), and hyperduplex (S32707 and S33207) are produced [16,17]. Superduplex (UNS S32750) is a highly alloyed grade having good chloride resistance and high mechanical strength. Recently, hyperduplex grade steels with higher amounts of alloying elements and N content up to 0.5 wt% are particularly suitable for use in severe marine solutions. The high Cr, Mo and N contents render them with excellent corrosion resistance, high strength and good formability for extrusion into seamless tubes for subsea umbilical applications [17]. Table 1 lists typical chemical compositions of commercial DSSs. The chemical compositions of DSSs play a dominant role in controlling their microstructures and properties. The processing, property and microstructure of DSS is rather complex. A comprehensive review on the processing-structural property of DSSs can be found elsewhere [1,2,3,4,5,6], and beyond the scope of this article.

Table 1: Chemical compositions (wt%) of duplex stainless steels (Fe content: balance)

UNS No.	EN No.	Common Name	C, max	Cr	Ni	Mo	Mn	Si	Co	Cu	N
S31803	1.4462	2205	0.03	21–23	4.5–6.5	2.5–3.5	2.0	1.0	–	–	0.08–0.2
S32205	1.4462	2205	0.03	22–23	4.5–6.5	3.0–3.5	2.0	1.0	–	–	0.14–0.2
S32550	1.4507	255	0.04	24–27	4.5–6.5	2.9–3.9	1.5	1.0	–	1.5–2.5	0.10–0.25
S32950	–	7Mo Plus	0.03	26–29	3.5–5.2	1.0–2.5	2.0	0.6	–	–	0.15–0.35
S32750	1.4410	2507	0.03	24–26	6.0–8.0	3.0–5.0	1.2	0.8	–	0.5	0.24–0.32
S32707	–	SAF 2707HD	0.03	27	6.5	4.8	1.0	0.3	1.0	–	0.4
S33207	–	SAF 3207HD	0.03	32	7	3.5	1.0	0.3	–	–	0.5
S32304	1.4362	2304	0.03	21.5–24.5	3.0–5.5	0.05–0.6	2.50	1.0	–	0.05–0.6	0.05–0.2
S32101	1.4162	LDX 2101	0.04	21–22	1.35–1.7	0.1–0.8	4.0–6.0	1.0	–	0.1–0.8	0.2–0.25

DSSs undergo microstructural changes during heat treatment, welding process and prolonged engineering service at high temperatures. Several undesirable precipitates such as carbides, nitrides, intermetallic phases (sigma and chi), and Cr-rich α' phase can induce in DSSs upon exposure to 950–400 °C temperature range [1,2,3,4,5,6,16]. Sigma phase is preferentially nucleated at the ferrite-austenite and ferrite-ferrite boundaries of DSSs at 700–950 °C [18,19]. Is formation depletes Cr and Mo at these regions, resulting in intergranular corrosion (IGC) and pitting corrosion [18,19,20,21,22,23]. Chromium carbides ($Cr_{23}C_6$) may form at the DSSs' grain boundaries upon heating at ~550–750 °C. The risk of $Cr_{23}C_6$ precipitation in commercial DSSs is rather small since their C content is kept at a maximum value of 0.03 wt%. At lower temperature regime of 350–550 °C, the α- phase transforms into Cr-rich (α') and Cr-poor (α) phases via a spinodal decomposition [24]. The Cr-rich α' precipitates reduces the impact toughness and corrosion resistance of DSSs significantly [25,26]. The spinodal decomposition is more pronounced at 475 °C, commonly referred to as the 475 °C embrittlement. This article focuses exclusively on the deleterious effect of these undesired phases on the corrosion resistance of DSSs.

SECONDARY PHASE FORMATION

Sigma, Chi and Nitride Phases

The literatures from 1983 to 2009 have reported the structures, morphologies and techniques for detecting intermetallic compounds and spinodal decomposed phases in details [1,2,3,4,5,6], with particular emphasis on the formation, characteristics and morphology of such precipitates and their adverse effects on the mechanical properties of DSSs [5,6]. The readers may refer to the literatures for further details [1,2,3,4,5,6]. Thus, we briefly presented such topics herein. Sigma phase exhibits tetragonal structure containing about 30% Cr, 4% Ni and 7% Mo [2]. The austenitic phase of DSSs exhibits closed-packed, face centered cubic structure. The precipitation reaction of σ-phase in the γ-phase is sluggish due to a slow diffusivity of solute atoms in this phase. As just mentioned, Cr, Mo and Si are ferrite formers that promote the σ-phase precipitation in DSSs aged at

700–950 °C. The diffusion rates of these elements in ferrite are much faster than in the austenite. As the precipitation continues, Cr and Mo diffuse to the σ-phase, leading to a depletion of these elements in ferrite, especially Mo content. Therefore, Mo from the inner region of ferrite diffuses to the σ-phase. Sigma phase nucleates preferentially at the α- and α-γ boundaries, then grows into the adjacent ferritic grains. Mo is the main element controlling secondary phase precipitation. It is noted that the χ-phase (bcc; $Fe_{36}Cr_{12}Mo_{10}$) containing about 25% Cr, 3% Ni and 14% Mo also forms between 700 and 900 °C, but in much smaller amounts [2]. The χ-phase contains even more Mo than the σ-phase, and nucleates in the early stage of aging due to the low interfacial energy of highly coherent χ/α interface with a characteristic of cubic-to-cubic orientation relationship [2,27]. Generally, ferrite can transform to secondary phases via the following eutectoid reaction,

$$\alpha \text{ (ferrite)} \rightarrow \sigma + \chi + Cr_2N$$

(1)

Although the χ-phase forms earlier than the σ-phase, it transforms into the σ-phase after prolonged aging [27,28]. The growth of σ-and χ-phase precipitates further depletes Cr and Mo in ferrite. Consequently, ferrite phase with high Ni content becomes unstable and eventually transforms into secondary austenite (γ_2). DSSs generally contain N up to 0.5 wt% for enhancing mechanical strength and pitting corrosion resistance (Table 1) Chromium nitride (Cr_2N; hexagonal structure) can form in DSSs with higher N content, and in the fusion zone of DSSs during welding.

Time-temperature transformation (TTT) diagrams obtained from isothermal heat treatment followed by quenching, can be used to investigate the susceptibility of DSSs to the σ-phase formation. This diagram indicates the time for a phase to decompose into other phases isothermally at different temperatures. In contrast, continuous-cooling transformation (CCT) diagram displays the time for a phase to decompose into other phases continuously at different rates of cooling. Figure 1 shows typical TTT and CCT curves for DSS 2205 with different σ-phase contents [29]. Isothermal heat treatment at 865 °C for 134 s precipitates 1% σ-phase. This implies that the aging time at 865 °C should not exceed 134 s. The 1% σ content can be obtained at room temperature after continuous cooling with 0.23 °C/s at 940 °C and above. To avoid the formation of more than 1% σ, the cooling rate from the solution annealing temperature must exceed 0.23 °C/s.

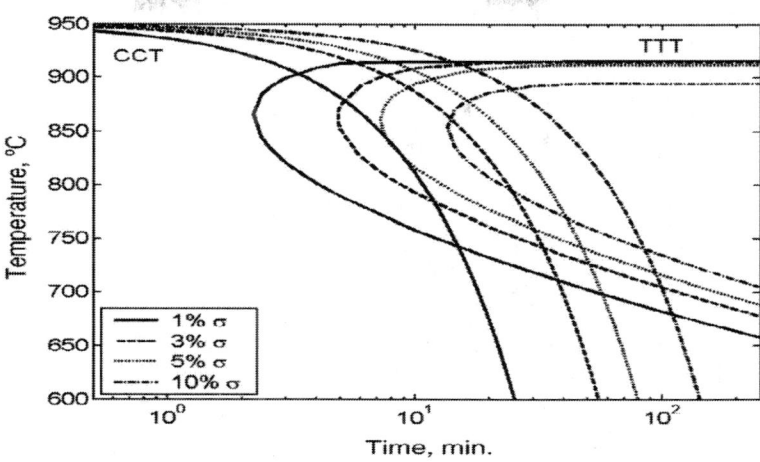

Figure 1: Time-temperature transformation (TTT) and continuous-cooling transformation (CCT) curves of duplex stainless steel (DSS) 2205 showing transformation of 1, 3, 5 and 10% σ-phase. The cooling rates are 0.23, 0.11, 0.07 and 0.04 °C/s, Reprinted with permission from [30].

Materials examination techniques are very important for the detection and identification of deleterious precipitates in DSSs. The structures of secondary phases can be determined with the X-ray diffraction (XRD), electron diffraction and electron backscattered diffraction (EBSD) techniques. The microstructural features of these phases in DSSs can be obtained using optical microscopy, scanning electron microscopy (SEM) and transmission electron microscopy (TEM) [27,30,31]. Combining with electron diffraction, TEM permits the examination of structural and morphological of localized region in a thin foil specimen at high magnifications. TEM is useful for characterizing the structures of fine precipitates such as σ- and χ- phases, especially in small amounts. However, the preparation of TEM thin film specimens is tedious, time consuming and difficult. EBSD involves the analysis of Kikuchi line bands formed by diffusively scattered electrons of individual crystals of a specimen. It provides easier identification of the crystal structure, grain orientation of unknown phases without the need for thin film sample preparation and enables observation of larger areas of a specimen. Recently, EBSD has been found to be very effective for detecting σ-phase formed in DSSs [32].

Figure 2 shows the XRD patterns of solution-annealed and aged DSS 2205 specimens. Solution annealing DSSs at high temperatures followed by water quenching is very effective to eliminate the σ-phase formation. The σ-phase peaks are observed in DSSs after aging at 800 and 900 °C. The typical solution temperature range is 1050–1080 °C. Above 1080 °C, e.g., at 1200 °C, unfavorable high α (62 vol%) and low γ-phase (38 vol%) content is produced [33]. The cooling rate from solution annealing temperature must exceed 0.23 °C/s to avoid the σ-phase formation as mentioned previously. The σ-phase with higher Cr and Mo contents than α- and γ-phases appears brighter in the back-scattered electron (BSE) image (Figure 3b) [31]. This figure reveals that the σ-phase forms preferentially at the α/γ and α/α boundaries, and penetrates into the α-phase. This phase can be clearly seen in the EBSD phase maps of aged DSS 2205, particularly after aging at 750 °C for 5 h (Figure 4) [32].

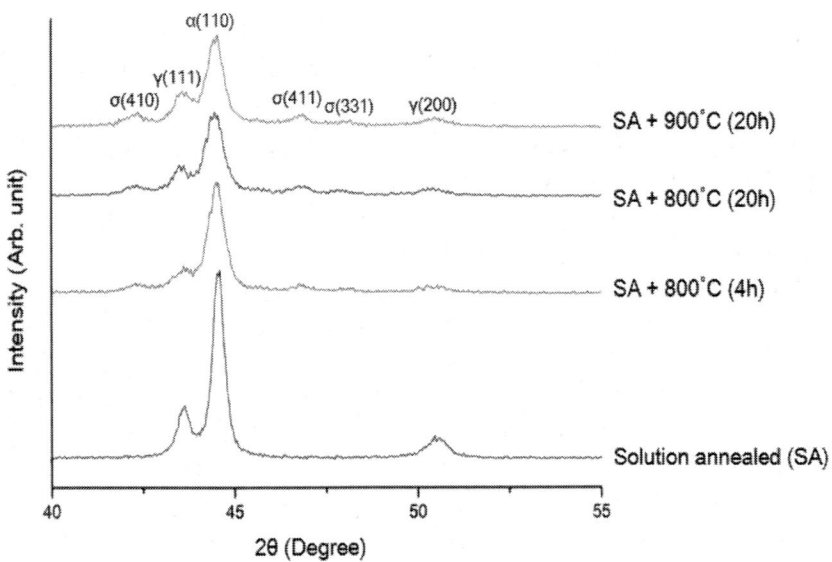

Figure 2: XRD patterns of solution-annealed DSS 2205, 4 h aged at 800 °C, 20 h aged at 800 °C and 20 h aged at 900 °C steel samples

(CuK radiation, wavelength: 0.154 nm).

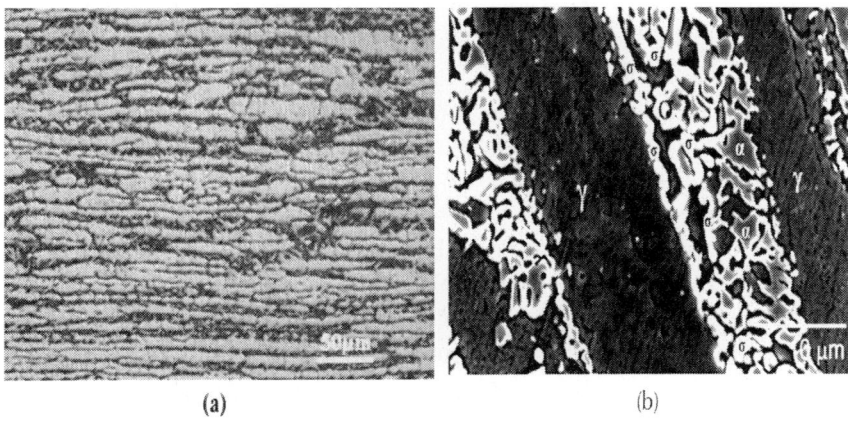

(a)   (b)

Figure 3: (a) Optical micrograph of DSS 2205 aged at 800 °C for 20 h. Dark phase is ferrite and bright phase is austenite; (b) Backscattered electron image of DSS 2205 aged at 875 °C for 20 min, reprinted with permission from [31].

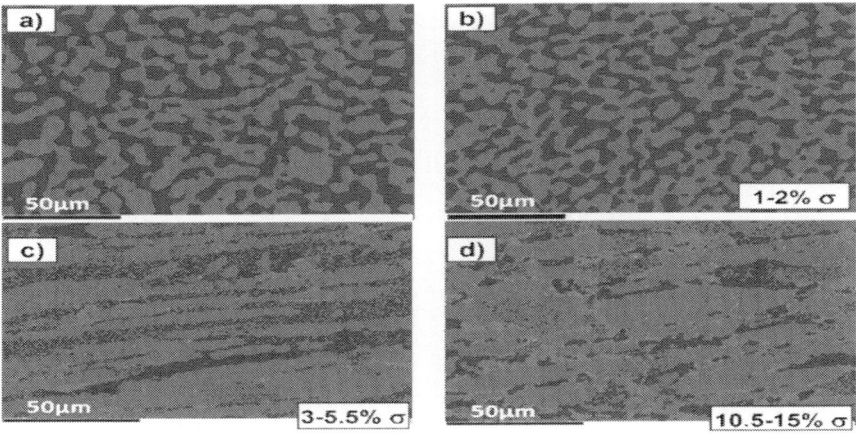

Figure 4: Electron backscattered diffraction (EBSD) phase maps for DSS 2205: (a) as-received; (b) aged at 900 °C for 30 min; (c) aged at 750 °C for 2 h and (d) aged at 750 °C for 5 h. Ferrite (blue), austenite (red), sigma (green), reprinted with permission from [32].

As mentioned earlier, bcc χ-phase forms in the earlier stage in the bcc ferrite of DSSs due to preferential cube-on-cube orientation relationship. The χ-phase then transforms to the σ-phase upon prolonged aging [27]. Figure 5a is the TEM image showing precipitation of χ-phase at the α-γ interface of DSS 2205 aged at 750 °C for 10 min only. The selected area electron diffraction (SAED) pattern and its index diagram are shown in Figure 5b,c, respectively. Michalska and Sozanska [31] reported that the χ and σ-phases precipitate preferentially at the α-γ interface and within the α-phase. The volume fraction and size of σ-phase increase with aging time. EDX results revealed that the χ- phase is highly enriched with Mo (15.95 wt%). Figure 6a shows the TEM image of DSS 2205 aged at 750 °C for 5 h. The SAED pattern is shown in Figure 6b. The σ-phase grows further into the α-phase after aging for 5 h because of high diffusivity of solute atoms at high temperatures. This phase is detrimental to the impact toughness of DSS 2205 (Figure 7).

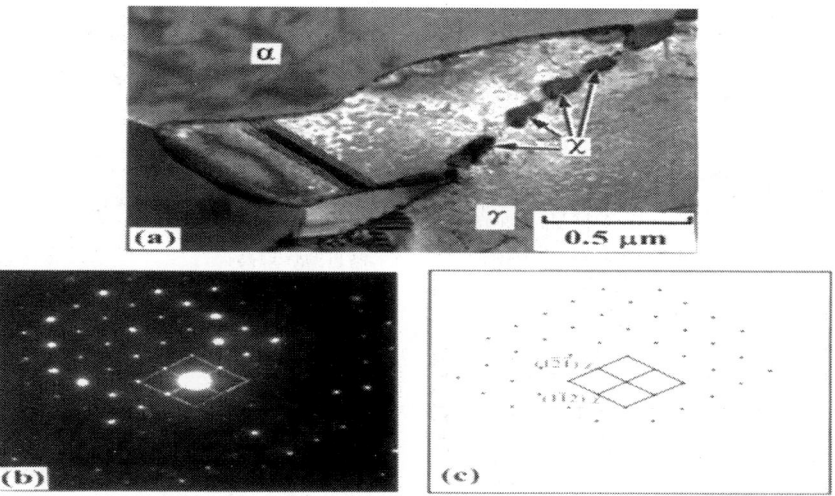

Figure 5: (a) TEM image of χ-phase of DSS 2205 aged at 750 °C for 10 min; (b) selected area electron diffraction (SAED) pattern with [5 3 1] zone and (c) index diagram, reprinted with permission from [27].

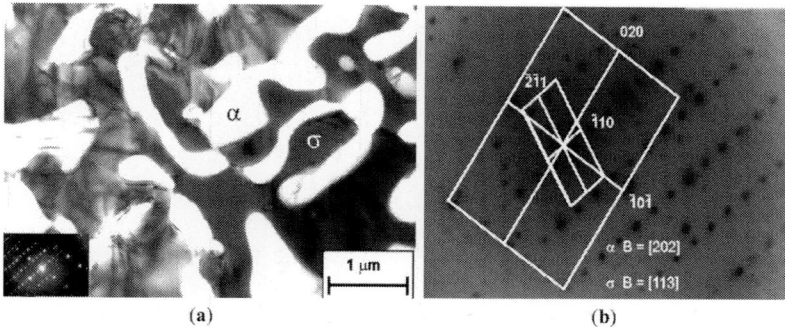

Figure 6: (a) TEM image showing σ-particle embedded in the α-phase matrix; (b) SAED pattern showing $[202]_\alpha$ and $[113]_\sigma$ zones, reprinted with permission from [31].

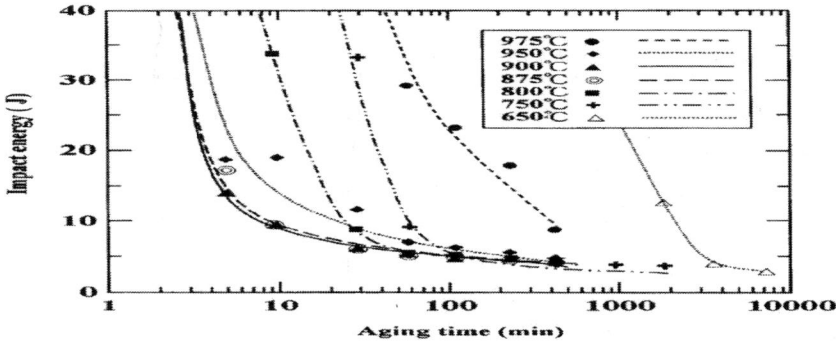

Figure 7: Effect of aging treatment on Charpy impact toughness of DSS 2205, reprinted with permission from [27].

For lean DSS S32101, the kinetics of σ-phase precipitation is much slower than S31803 due to its reduced Cr and extremely low Mo contents. Fine Cr_2N particles nucleate at the α/α and α/γ boundaries by aging at 700 °C at the initial stage of aging (240 min) [34,35,36]. The Cr_2N precipitation depletes Cr in the ferrite phase, resulting in the γ_2 formation via a reaction $\alpha \rightarrow Cr_2N + \gamma_2$. Further aging to 168 h leads to the formation of a small amount of σ-phase adjacent to the Cr_2N precipitates (Figure 8). Prolonged aging for 300 h yields an increase in the σ-phase volume fraction.

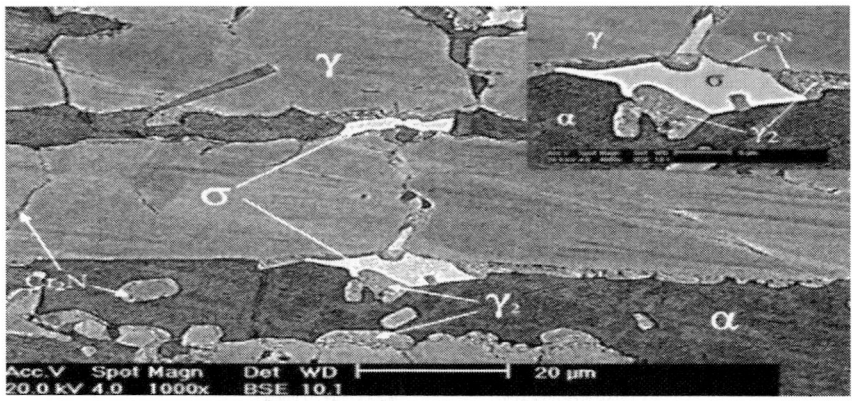

Figure 8: Backscattered electron image of UNS S32101 steel aged at 700 °C for 168 h, reprinted with permission from [36].

Spinodal Decomposition

Spinodal decomposition of ferrite is originally found in ferritic Fe-Cr alloys heated at 280–500 °C due to the presence of a miscibility gap in the alloy system. Duplex stainless steels also suffer from the "475 °C embrittlement" associated with the precipitation of Cr-rich '-phase and Fe-rich α-phase within ferritic grains. This decomposition occurs at 350–550 °C, with the most rapid formation at 475 °C [37,38,39]. Weng et al. [38] aged DSS 2205 at 400–500 °C for different times, and examined microstructures and mechanical properties of the aged specimens. Low-temperature aging treatment causes a significant reduction in the impact strength of DSS 2205, especially at 475 °C with aging times ≥1000 min (Figure 9). They also reported that the ferrite phase exhibits a modulated contrast, displaying the appearance of an orange-peel because of aging treatment. The mottled aspect is caused by compositional fluctuations associated with the formation of Cr-rich and Cr-poor phases. Furthermore, highly dense dislocations are created in the ferrite phase owing to a difference in thermal expansion coefficient between ferritic and austenitic grains upon cooling from the decomposition temperatures. The immobilization of dislocations hardens ferritic grains considerably (Figure 10) [37,38,39,40]. Prolonged aging treatment leads to the formation of G-phase (fcc; 25% Cr, 25% Ni, 4% Mo) along the dislocations in the ferrite phase (Figure

11) [2,41]. It is noted that TEM mottled contrast gives misleading indication for spinodally decomposed phases in DSSs. This is because the difference in lattice parameter between the Cr-rich and Fe-rich is very small. The Cr-rich phase is coherent with the alloy matrix. There is no clear interface between the precipitate and the matrix. Thus it is difficult to detect accurately the composition fluctuations in spinodally decomposed products using electron microscopy. However, atom probe technique is attractive to study the ultrafine scale phase separation and compositions in sub-nanometer scale. Some researchers used the atom probe technique in combination with field ion microscopy to elucidate this issue [42,43,44,45,46].

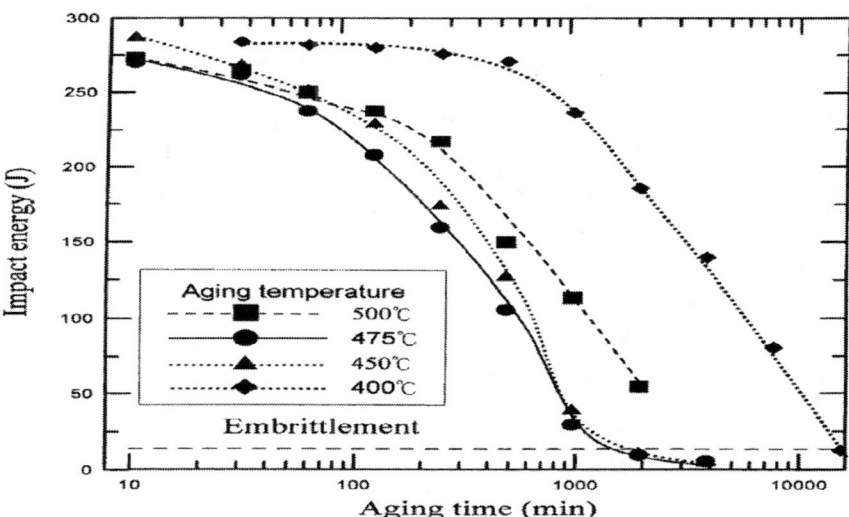

Figure 9: Impact energy *vs.* aging time for DSS 2205 treated at 400–500 °C, reprinted with permission from [38].

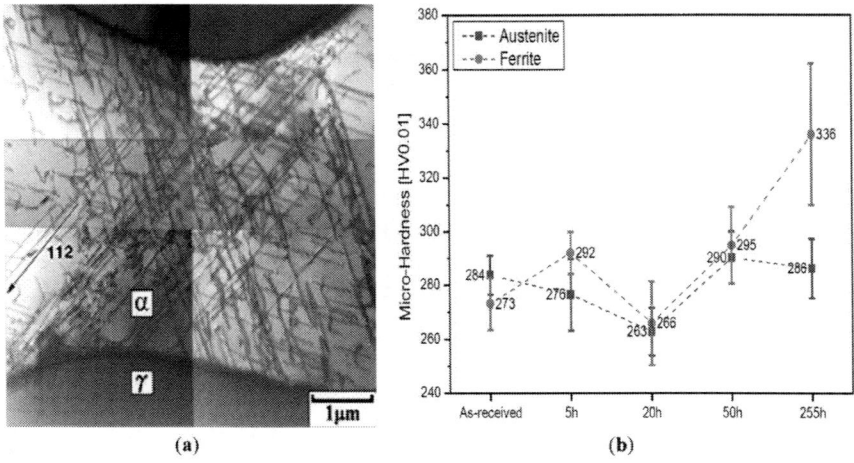

(a)   (b)

Figure 10: (a) Dislocation structure in DSS 2205 aged at 475 °C for 64 h, reprinted with permission from [38]. Copyright 2004 Elsevier; (b) Microhardness of ferritic and austenitic grains *vs.* aging time of DSS 2205 treated at 475 °C, reprinted with permission from [37].

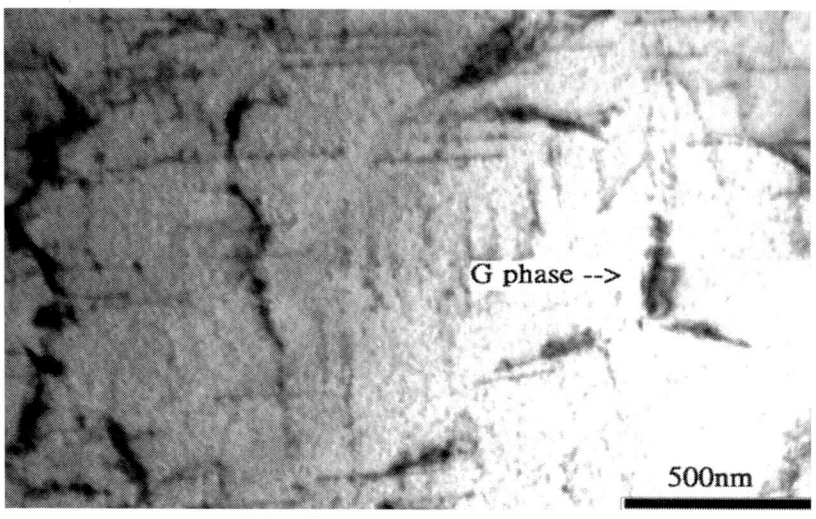

Figure 11: Development of dislocations and precipitation of G-phase in the 7MoPlus steel aged at 500 °C for 1744 h, reprinted with permission from [41].

Recently, Lo and Lai [41] demonstrated that a.c. magnetic susceptibility is an effective tool for detecting spinodal decomposition of the 7MoPlus steel (UNS S32950). Figure 12 shows magnetic susceptibility *vs.* aging time for 7MoPlus. Magnetic susceptibility is a dimensionless quantity that describes the magnetization of a material in response to an applied magnetic field. Apparently, magnetic susceptibility of the specimens aged at 450–550 °C decreases markedly with aging time up to 1000 h, then gradually approaches a steady state value with increasing time. At 475 °C, the reduction in susceptibility is more pronounced. The marked reduction in magnetic susceptibility at the earlier stage of aging is due to the decomposition of ferromagnetic primary ferrite to paramagnetic Cr-rich '-phase and ferromagnetic Fe-rich -phase. The magnetic susceptibility of aged specimens does not diminish to zero to the formation of ferromagnetic Fe-rich phase. Moreover, primary ferrite that does not fully undergo spinodal decomposition also contributes to the susceptibility. In contrast, magnetic susceptibility of the steel aged at 350 °C remains nearly unchanged, indicating no spinodal decomposition of primary ferrite during aging.

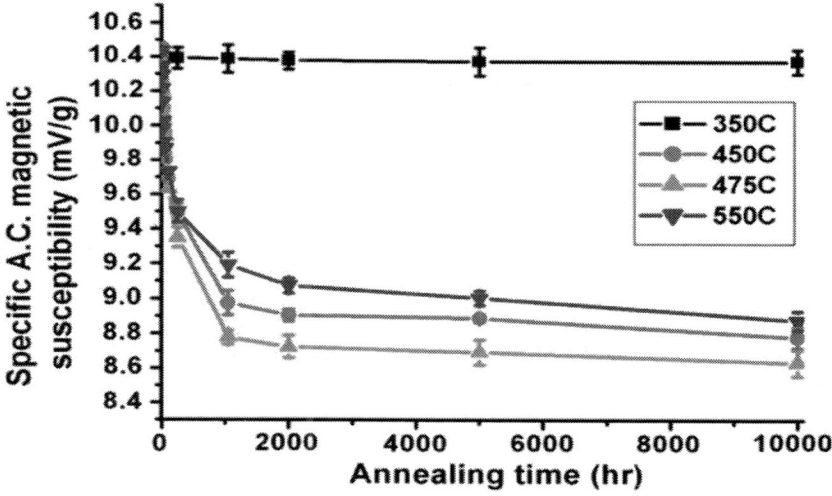

Figure 12: Variation of magnetic susceptibility with aging time of 7Mo-Plus steel treated at 350, 450, 475 and 550 °C, reprinted with permission from [41].

CORROSION PROPERTIES

As recognized, protective oxides or hydroxides can form on the surfaces of transition metals such as Fe, Cr, Ni, Mo and their alloys. The oxide film with a thickness of several nanometers protects underlying metal/alloy from corrosive environment, commonly terming as the passive film. Therefore, annealed DSSs are well protected by the passive films formed on their surfaces. The protectiveness of passive film depends greatly on the Cr, Mo and N contents. In particular, Cr and Mo in the passive film act synergistically in resisting the attack of chloride ions by rehealing damaged film. When secondary phase particles and chromium carbides are formed at the grain boundaries of DSSs, the boundaries adjacent to the precipitates are depleted of Cr and Mo. The Cr/Mo-depleted zone near the grain boundaries is much less corrosion resistant than the surrounding grains. Thus the film locally is less protective and the Cr/Mo depleted zone experiences active dissolution (act as the anode) and corrode upon exposure to corrosive environment, while the surrounding grains remain in the passive state (act as the cathode). The active-passive behavior of metals in aqueous solutions can be determined from a plot of applied potential *versus* current density using a potentiostat. This instrument gives a continuously varying potential to the specimen. As the applied potential is varied, the current is continually recorded. The resulting current density-potential plot is known as the polarization curve. At the passivity domain, the current density of metal dissolution decreases drastically associated with the formation of a stable passive film.

DSSs exhibit superior corrosion resistance than austenitics in acidic and marine environments. The degree of corrosion protection increases with increasing Cr, Mo and N contents. In addition, DSSs also perform satisfactorily in caustic solutions. Recently, Bhattacharya and Singh investigated corrosion behaviors of the as-received S32205, S32101 and S32304 in both 3.75 M NaOH and 3.75 M NaOH + 0.64 M Na_2S solutions at 40–170 °C [47]. Sodium sulfide is added because the pulp mill facilities always contain sulfide species. All DSSs exhibit good passivation behavior in the 3.75 M NaOH solution (Figure 13). In this figure, E_{corr} of S32205 is located at −1.04 V (SCE). For the potentials more cathodic than E_{corr}, hydrogen ion reduction reaction takes place as expected. The primary passive region extends from about −1.1 V to −0.3 V with a low current density of ~10^{-5} A/cm^2. −0.3 to 0 V potential range, the anodic current increases due to the transpassive oxidation of Cr. Transpassive oxidation of a metal is defined as the formation of chemical species in a valence state higher than that in the primary passive film formed on the material. In other words, Cr^{III} species in the passive film of DSS 2205 is further oxidized to Cr^{VI}. Above 0 V, secondary passivation occurs followed by the oxygen evolution. Thus, the anodic polarization of S2205 in NaOH solution is rather complex, consisting of primary passivation at lower anodic potential, followed by transpassive oxidation and secondary passivation at high anodic potentials [48]. Furthermore, temperature and Na_2S species affect general corrosion rates of DSSs in caustic environments (Figure 14). The corrosion rates of these steels in 3.75 M NaOH solution are below 0.2 mm/year even at 170 °C. At temperatures ≤90 °C, the rates are an order of magnitude smaller. In the presence of Na_2S, the passivation of DSSs in alkaline solution degrades considerably, leading to an increase in the critical current density for passivation and a reduction in the passivation range. Bhattacharya and Singh attributed this to the formation of metal-sulfide compounds in DSSs. Such sulfide compounds are less protective than passive films enriched with Cr-oxide/hydroxide. S32205 is more susceptible to general corrosion than lean DSSs because Mo in the S32205 undergoes active dissolution. Lean S32304 DSS exhibits the lowest corrosion rates in caustic sulfide environmenent due to its lowest Mo content of 0.2 wt%.

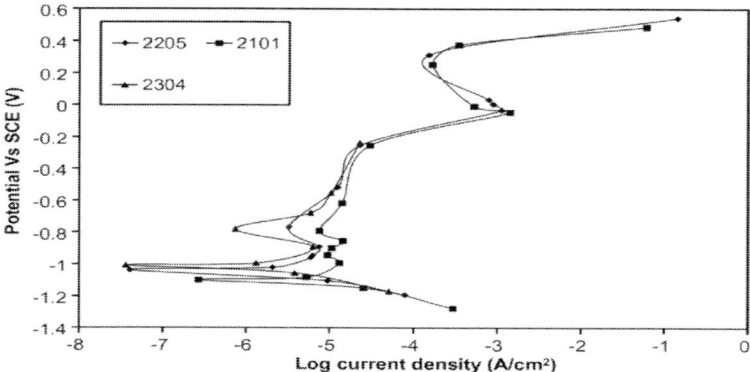

Figure 13: Polarization curves of UNS S32205, S32101 and S32304 steels in 3.75M NaOH solution at 70 °C, reprinted with permission from [47].

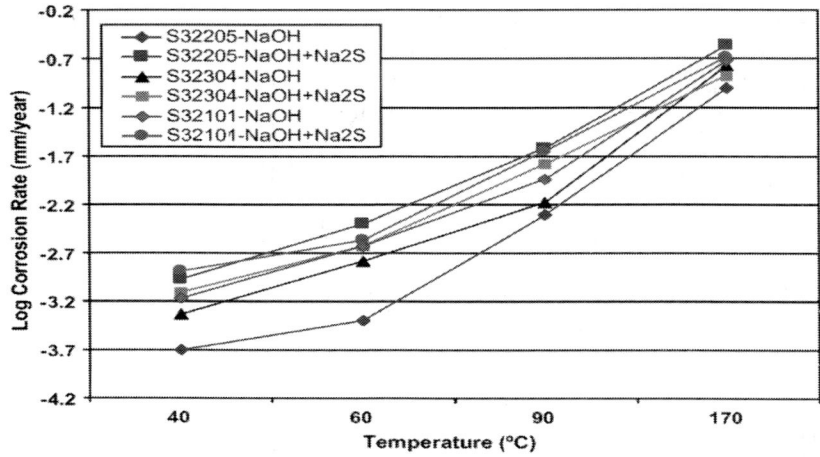

Figure 14: General corrosion rate as a function of temperature determined by immersing UNS S32205, S32101 and S32304 steels in caustic solutions with and without sulfide addition, reprinted with permission from [47].

Intergranular Corrosion

As mentioned above, the -phase formation impairs Charpy impact toughness of DSSs significantly (Figure 7). Moreover, the formation of intermetallic compounds, chromium nitrides and chromium carbides depletes Cr or both the Cr and Mo in DSSs, leading to intergranular corrosion upon exposure to corrosive environments. Accordingly, ASTM A923-03 standard was set up to detect detrimental intermetallic phases in duplex austenitic/ferritic stainless steel that lead to low toughness and poor corrosion resistance [49]. The practice includes method A-sodium hydroxide etch test, method B-Charpy impact test, and method C-ferric chloride corrosion test. Test method A is a screening test for methods B and C. Method C is an immersion weight loss test in ferric chloride solution for 24 h, showing the loss of corrosion resistance due to the depletion of Cr and Mo associated with the precipitation of Cr-rich and Mo-rich phases. From these, it appears that only method C is a quantitative practice for evaluating susceptibility of DSSs to intermetallic compounds. Performing of weight-loss measurements in corrosive media is destructive. Thus, it is necessary to develop other corrosion methods for detecting IGC.

Single loop electrochemical potentiokinetic reactivation (EPR) was originally developed for evaluating the degree of sensitization of austenitic stainless steels [50]. The EPR test is a quantitative method by measuring the amount of charge resulting from dissolution of Cr-depleted regions. In this test, sensitized specimen is first passivated in 0.5 M H_2SO_4 + 0.01 M KSCN solution at 0.2 V vs. saturated calomel electrode (SCE) for 2 min. This is followed by scanning the potential towards active direction at a rate of 6 V/h, down to the E_{corr}. The area under the reactivation peak (charge Q) is normalized by the grain boundary area (grain size), reflecting the degree of sensitization (DOS). In contrast, unsensitized stainless steel yields no reactivation peak. Single EPR test requires fine surface finish (1 μm) and grain size determination for detecting DOS [51].

Majidi and Streicher [52] modified the test by polarizing austenitic UNS S30400 in 0.5 M H_2SO_4 + 0.01 M KSCN solution from open-circuit potential to the passive region with a constant scan rate. Subsequently, the scan is reversed towards active potential region, i.e., open-circuit potential with the same scan rate. The DOS is defined by the I_r/I_a × 100, where I_a is maximum current density in forward scan (activation) and I_r is the peak current density in reverse scan (reactivation) as shown in Figure 15. On the reverse scan, the passive film formed on Cr-depleted regions during the forward scan degrades considerably, leading to corrosion attack in these areas and generating current peak I_r. This procedure becomes the basis of the ISO 12732 practice [53], and widely known as double loop electrochemical potentiokinetic reactivation (DL-EPR) test. Microstructural factors such as the grain size and surface finish are not accounted for the DL-EPR test. Thus grinding the samples in fine #600 grit SiC paper is commonly adopted. It offers an opportunity for nondestructive measurement of the DOS of stainless steels of different grades with great simplicity [54].

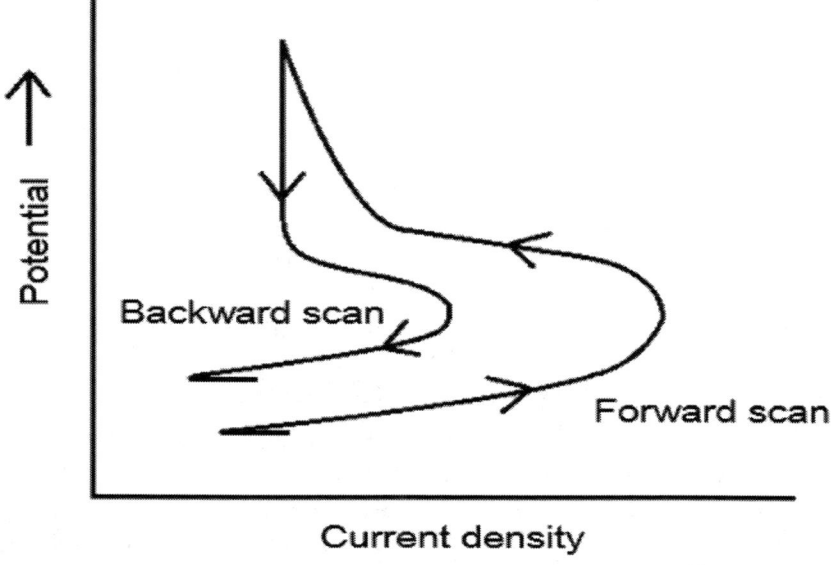

↑ Potential

Backward scan

Forward scan

Current density

Figure 15: Schematic diagram showing double loop electrochemical poten-tiokinetic reactivation (EPR) test.

Lopez *et al.* [18] employed DL-EPR test to study the effect of -phase formation on the IGC of UNS S31803 aged at 600–900 °C for different times. The test solution was modified to 2 M H_2SO_4 + 0.01 M KSCN + 0.5 M NaCl solution at 30 ± 1 °C. KSCN is less effective to de-passivate DSS with better corrosion resistance, thus NaCl acting as a depassivator is added simultaneously. The scan rate used was 1.66 mV/s. They reported that DL-EPR test is effective for detecting DOS resulting from the dissolution of Cr-depleted regions. The DL-EPR results agree closely with those of oxalic acid etch test and corrosion rate measurement (immersion test). Very recently, Ortiz *et al.* [55] also evaluated IGC susceptibility of UNS S31803 in 2 M H_2SO_4 + 0.01 M KSCN + 0.5 M NaCl solution at room temperature using DL-EPR method at a scan rate of 1 mV/s. The steel specimens were aged at 700 °C from 1 min to 240 h. The DOS values of all specimens are listed in Table 2. Apparently, the as-received steel and specimens aged for times ≤1 h exhibit very small DOS values, demonstrating no IGC attack. By increasing aging times to 6 h and above, the DOS values increase to about three orders of magnitude higher due to the precipitation of σ-phase.

Table 2: Degree of sensitization (DOS) values of as-received UNS S31803 steel and DSS specimens aged at 700 °C for different times, reprinted with permission from [55]

Specimen	Activation Peak Current Density, I_a (A/cm²)	Reactivation Peak Current Density, I_r (A/cm²)	DOS $(I_r/I_a) \times$ 100%
As received	0.0163	1.59×10^{-6}	9.74×10^{-3}
1 min	0.0126	2.63×10^{-6}	2.09×10^{-2}
30 min	0.0177	4.16×10^{-6}	2.34×10^{-2}
1 h	0.0066	5.06×10^{-6}	7.66×10^{-2}
6 h	0.0240	0.0041	17.40
12 h	0.0251	0.0083	33.15
24 h	0.0472	0.0228	48.37
48 h	0.0579	0.0376	64.91
120 h	0.0830	0.0579	69.69
240 h	0.0753	0.0669	88.83

To improve sensitivity of DL-EPR, HCl depassivator of different concentrations, electrolyte temperatures and scan rates were investigated by some researchers [23,36,56,57,58]. Hong et al. [23] reported that the optimal conditions for evaluating DOS of S32750 in a 2 M H_2SO_4 solution can be obtained by using 1.5 M HCl, a scan rate of 1.5 mV/s and a solution temperature of 30 °C. Gong et al. [57] reported that an electrolyte of 2 M H_2SO_4 + 1 M HCl at 30 °C and a scan rate of 1.66 mV/s give optimal DL-EPR responses for aged UNS S31803. Figure 16a shows the DOS vs. aging time for S31803 determined under these conditions. The DOS reaches an apparent maximum value of 64.8% by aging at 800 °C for 24 h, and then decreases to 42.85% as the aging time increases to 48 h. The decrease in DOS value is attributed to the self-healing of Cr-depleted zones associated with the diffusion of Cr and Mo from the matrix to depleted regions. Such self-healing behavior also occurs in S32750 [23]. For lean UNS S32101 DSS, an electrolyte of 33% H_2SO_4 + 0.1% HCl at 20 °C and a scan rate of 2.5 mV/s give the best reproducibility [36]. The DOS values vs. aging time are shown in Figure 16b. Apparently, the DOS values increase markedly with increasing aging time. The initial small increase in DOS value is caused by the depletion of Cr associated with the Cr_2N precipitation as mentioned previously. The IGC becomes more serious once the -phase begins to nucleate in this steel (Figure 8), producing large DOS value of 31.5% as expected. There is no self-healing in lean DSS after aging up to 300 h. Amadou et al. [56] studied the applicability of the DL-EPR method for detecting intergranular corrosion susceptibility of UNS S31260 and S31803 steels aged at 500 °C to 900 °C for different periods ranging from 6 min to 120 h. The electrolyte and scan rate used were 33% H_2SO_4 + 0.3% HCl (room temperature) and 2.5 mV/s. A wide range of aging temperatures was selected in order to induce precipitation of $Cr_{23}C_6$ carbides (550–650 °C) and intermetallic phases (650–850 °C). DOS was also defined from the ratio of reactivation charge (Q_r) and activation charge (Q_a) given by the relation $DOS = Q_r/Q_a$. Solution annealed steel exhibits DOS value with $Q_r/Q_a < 1$, or $I_r/I_a < 1$. In contrast, sensitized steel displays the $Q_r/Q_a \geq 1$, or $I_r/I_a \geq 1$ behavior. They demonstrated that the DL-EPR technique is very effective to detect IGC caused by the Cr-depletion zone due to the precipitation of $Cr_{23}C_6$ carbides, and phases in DSSs.

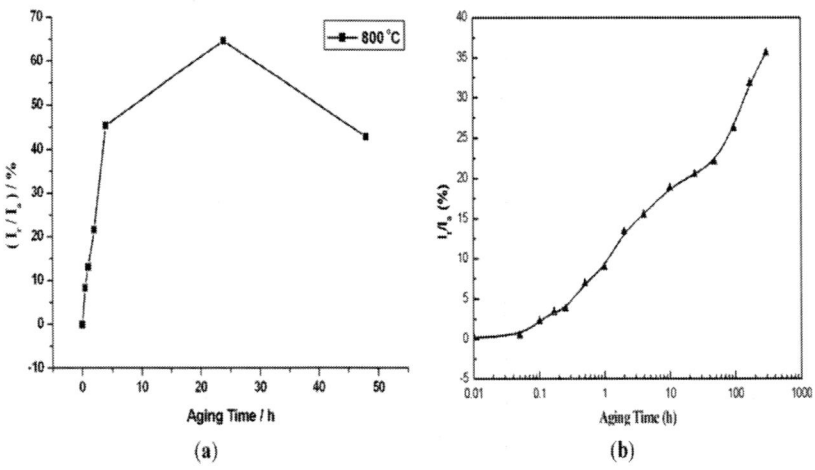

Figure 16: DOS vs. aging time of (a) UNS S31803 steel treated at 800 °C, reprinted with permission from [57]. Copyright 2010 Elsevier; and (b) UNS S32101 steel treated at 700 °C, reprinted with permission from [36].

It is worth noting that DL-EPR test can also be used to characterize the Cr depletion in DSSs due to the formation of Cr-rich ʹ-precipitates during the spinodal decomposition [26]. Figure 17 shows the plots of $(I_r/I_a \times 100)$ vs. aging time for the 7MoPLUS steel aged at 300–500 °C for extended periods. At low temperatures of 300 and 400 °C, the $(I_r/I_a \times 100)$ values of aged specimens remain relatively low even after prolonged aging for 15,000 h. However, the $(I_r/I_a \times 100)$ value increases significantly at 500 °C due to the formation of Cr-depletion zone next to the Cr-rich ʹ-phase.

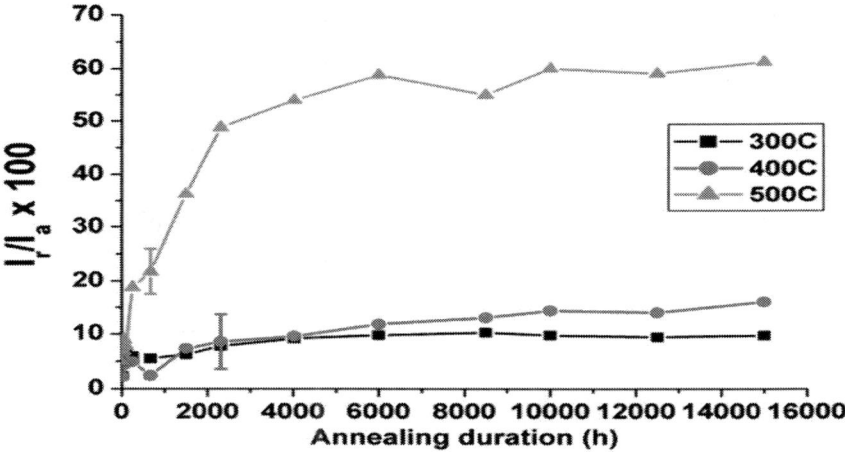

Figure 17: $(I_r/I_a \times 100)$ value vs. aging time plots of 7MoPLUS steel aged at 300–500 °C, reprinted with permission from [26].

From these results, it appears that DL-EPR can be used to detect the Cr depletion zone resulting from the precipitation of $Cr_{23}C_6$ carbides, intermetallic compounds and Cr-rich ʹ-precipitates. Therefore, the reproducibility of the test is considered of technological importance. The sensitivity of DL-EPR depends greatly on the operating condition (e.g., scan rate), electrolyte temperature and pH, depassivator content, etc. At present, there is no general agreement among researchers concerning the concentrations of electrolyte and depassivator, the scan rate and electrolyte temperature for this test [23,36,55,56,57,58]. The optimal conditions for achieving good reproducibility are still under investigation. Amadou et al. [56] determined the DOS from the ratio of reactivation charge and activation charge during the DL-EPR test, but the DOS value was not normalized by the grain size (grain boundary area) of duplex steels as the DOS in austenitic stainless steels. Therefore, careful consideration should be given to the issues of microstructural and electrolyte parameters such that the DL-EPR test can yield data with high reproducibility.

Localized Corrosion

It is well recognized that crevice and pitting corrosion are caused by a breakdown of the passive films of stainless steels exposed to chloride containing environments. Chloride ions can induce failure of the passive film formed on the inclusions (e.g., MnS) and secondary phases, leading to anodic dissolution of underlying metal locally. The adjacent metal surfaces act as the cathode for oxygen reduction process. Rapid dissolution at localized region initiates the pit formation. An excess of metallic ions with positive charge is accumulated in this region, resulting in the migration of chloride ions from the solution to maintain electroneutrality. Consequently, a high concentration of MCl within the pit undergoes hydrolysis, producing low pH due to the formation of hydrochloric acid. This increases the local dissolution rate causing more chloride ions to migrate into the pit. The process is a self-propagating or autocatalytic mechanism of pit growth [59]. Moreover, pitting can lead to other causes of failure such as stress corrosion. The simultaneous presence of tensile stress and corrosive environment can breakdown the passive film even more readily, forming fine cracks that fracture in a brittle manner. These cracks propagate across the grains in either transgranular or intergranular mode. This phenomenon is referred to as the "stress corrosion cracking" (SCC). Residual or thermal stresses resulting from the fabrication, improper heat treatment and welding process of DSSs form the basis for tensile loads. Moreover, the presence of intermetallic phases at the α/α and α/γ grain boundaries facilities the cracks to propagate along those boundaries.

Pitting Corrosion

The pitting resistance of DSSs in marine environment depends mainly on their Cr, Mo and N contents. Thus, the pitting corrosion resistance is correlated with the chemical compositions of DSSs, and can be expressed in term of an empirical pitting resistance equivalent number

(PREN) given by [12]:

$$PREN = wt\% \ Cr + 3.3 \ wt\% \ Mo + xwt\% \ N$$

(2)

where the x value ranges from 16 to 30. Larger PREN value gives rise to higher pitting resistance. There is no universal equation for evaluating PREN of DSSs. A value of $x = 16$ is typically adopted in industrial sector [16], while $x = 20$ is commonly used by the researchers [60,61,62]. For $x = 16$, the PREN values of standard 2205 grade, 25% Cr, superduplex and hyperduplex steels are ~ 35, 35–40, 40–50 and ≥50, respectively [16]. Equation (5) only takes into account the beneficial effect of Cr, Mo and N on the pitting corrosion resistance of DSSs. However, other elements such as Mn, S and P that exhibit deleterious effect on the pitting resistance are ignored. Partitioning of Cr and Mo in ferritic phase, and of Ni and N in austenitic phase can affect the PREN values of both phases. Furthermore, the precipitation of secondary phase particles can cause compositional changes in the - and -phases, resulting in selective pitting corrosion of a weak phase [22].

In addition to intrinsic material properties, environmental parameters such as chloride concentration, pH and temperature of the solutions also affect the pitting corrosion resistance of DSSs. Figure 18 shows the effect of temperature on polarization behavior of solution-annealed DSS 2205 in 0.1 M NaCl solution. Apparently, the current density of DSS in passivation region increases with increasing solution temperatures [63]. Moreover, the pitting potential of DSS shifts to less noble value as the solution temperature increases. Pitting potential in the polarization curve is characterized by the dramatic increase in current density at a critical potential. This potential is taken as a measure of resistance to pitting corrosion. The pitting potential of DSS 2205 in 0.1 M NaCl solution at 25–45 °C is 1050 mV, but shifts to more active value of 400 mV as the solution temperature increases to 65 °C. It is noted that the scans do not show a true pitting potential at 25–45 °C since the current increases into transpassivity at 1050 mV.

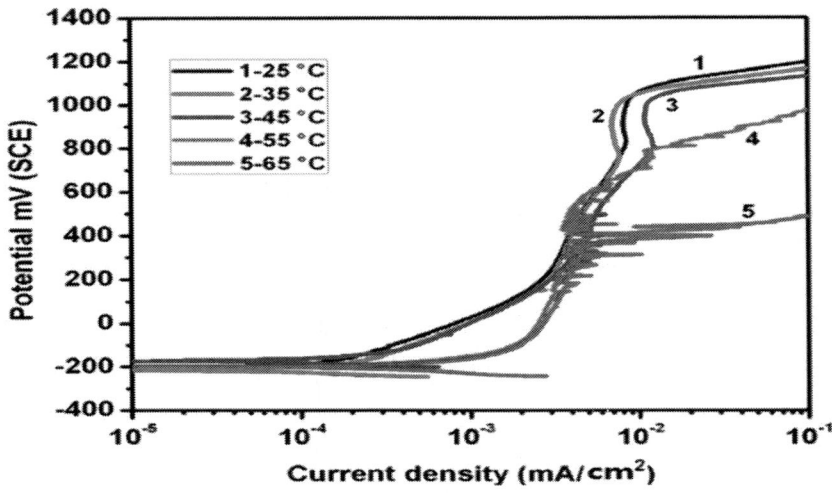

Figure 18: Polarization curves of 2205 DSS exposed in 0.1 M NaCl solution at different temperatures, reprinted with permission from [63].

Domínguez-Aguilar and Newman studied the deleterious effects of secondary phase formation on the pitting behavior of UNS S32760 containing 24.97 wt% Cr and 3.58 wt% Mo exposed in 0.85 M halide solutions (NaBr, NaCl) at different temperatures (20–50 °C) [64]. The steel was aged at 67–825 °C for different periods to induce the formation of   and   phases. As mentioned above, Mo promotes the formation of - and -phases in DSSs in which the -phase nucleates in the early stage of aging. In both halide solutions the presence of intermetallics leads to pitting corrosion. The Cr/Mo depleted zones are preferred dissolution sites in halide solutions. The passive films formed on these zones are less protective because they are depleted of Cr and Mo. The bromide solution is more effective to detect mild solute depletion at room temperature, while chloride solution at room temperature is ineffective and must be heated to 50 °C. Figure 19a shows the effects of aging temperature and time as well as -phase volume content on the pitting potential of S32760 exposed in halide solutions. The pitting potential of this steel immersed in 0.85 M NaCl solution (50 °C) and 0.85 M NaBr solution (room temperature) decreases with increasing -phase content as expected. The correlation between pitting potential and -phase volume fraction is shown in Figure 19b, which shows a shift of pitting potential to more active region as the -phase content increases.

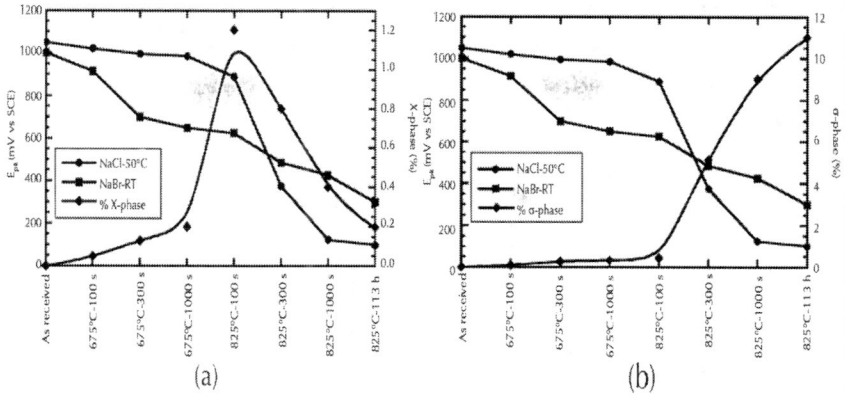

(a)   (b)

Figure 19: Variation of pitting potential of S32760 exposed in halide solutions with (a) χ-phase and (b) σ-phase volume fraction, reprinted with permission from [64].

As recognized, pitting potential determined from anodic polarization tests depends greatly on the scan rate and fine crevice formed at the sample/resin interface. For the sample mounted in a resin, the crevice effect lowers its pitting potential considerably. In this respect, critical pitting temperature (CPT) is more appropriate to describe the susceptibility of stainless alloys to pitting corrosion [65]. From the ASTM G150 test, CPT values of stainless steels can be determined using potentiostatic polarization [66]. In this test, an anodic potential of 700 mV *vs*. SCE is applied to the specimen immersed in 1 M NaCl solution. The current density is recorded while the solution temperature is continuously increased at a rate of 1 °C/min until stable pitting occurred. CPT is defined as the temperature when the sample current density reaches 100 µA/cm^2.

Li *et al*. [33,60] studied the effect of microstructural evolution on the CPT behavior of lean S32304 and superduplex S32750 steels by annealing at 1000–1200 °C for 1–2 h followed by water quenching. Annealing at high temperatures disturbs the balance of - and -phases in these steels. From the results of image analysis, the ferrite volume fraction increases while the austenite fraction decreases with increasing temperature. Energy dispersive X-ray measurements reveal that the Cr and Mo contents in the -phase decrease while N content in the -phase increases with increasing temperature. Accordingly, PREN value determined with $x = 20$ for the -phase decreases, while that for the -phase rises by increasing the temperature (Figure 20). At a cross-over temperature of 1080 °C, PREN values of both phases are almost the same, demonstrating that the - and -phases have equal pitting resistance. Consequently, pits are nucleated at / boundaries. Below 1080 °C, PREN value of -phase is smaller than that of -phase, implying that -phase is more corrosion resistant. Therefore, pits are preferentially nucleated in the -grains. Above 1080 °C, PREN value of -phase is smaller than that of -phase, thus pits are nucleated in the -phase accordingly. Figure 21 shows the application of potentiostatic polarization method for determining CPT of annealed UNS S32750 in 1 M NaCl solution. CPT values are determined from potentiostatic measurements by applying an anodic potential of 750 mV (SCE) to the specimen and continuously increasing solution temperature at a rate of 1 °C/min until stable pitting occurred. CPT is defined as the temperature at which the current density reaches 100 $\mu A/cm^2$. Treating S32750 at 1080 °C produces the highest CPT of 96 °C, *i.e.*, the steel exhibits the best pitting resistance at this temperature. The CPT *vs.* annealing temperature of S32750 is also plotted in Figure 20. The CPT value generally follows rise/fall PREN trend of less corrosion resistant phase. In other words, CPT first increases with annealing temperature up to 1080 °C and follows rising trend of austenitic PREN, then decreases with annealing temperature and follows falling trend of ferritic PREN. Table 3 summarizes the effect of annealing temperature on pitting susceptibility of S32750 and S32304 steels. Lean S32304 exhibits low CPT values in 1 M NaCl solution as expected due to its extremely low Mo (0.3 wt%) and lower Cr (22.9 wt%).

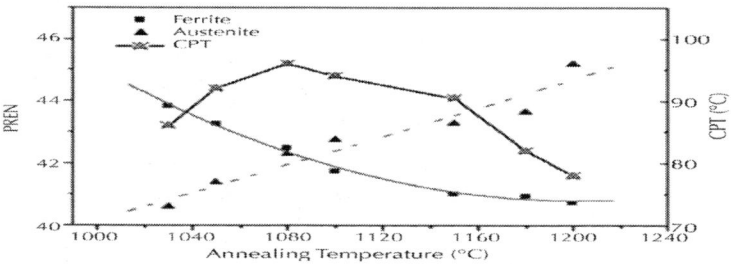

Figure 20: Variations of α- and γ-phase pitting resistance equivalent number (PREN) of UNS S32750 and its critical pitting temperature

(CPT) in 1 M NaCl solution with annealing temperature, reprinted with permission from [33].

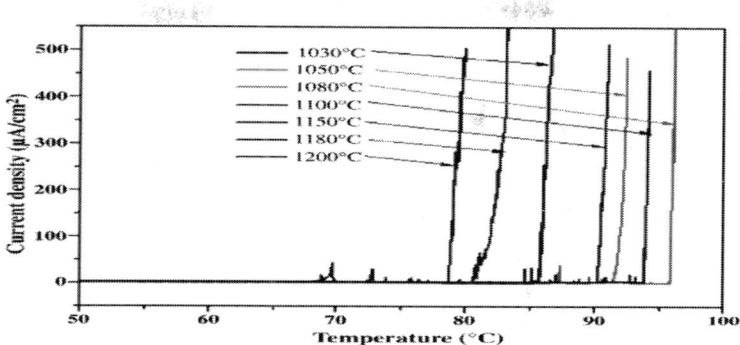

Figure 21: Potentiostatic CPT measurements of UNS S32750 in 1 M NaCl solution under a potential of 750 mV(SCE), reprinted with permission from [33].

Table 3: Effect of annealing temperature on pitting behavior of S32750 and S32304 steels [33,60]

Material	Annealing Temperature, °C and Time	-Phase Content, vol%	-Phase Content, vol%	CPT (°C), 1 M NaCl	Pitting Phase
S32750	1050, 2 h	47	53	92	
	1080, 2 h	49	51	96	/ boundary
	1150, 2 h	55	45	90	
	1200, 2 h	62	38	78	
S32304	1030, 1 h	47	53	58	
	1080, 2 h	51	49	67.8	/ boundary
	1150, 1 h	58	42	56	
	1200, 1 h	63	37	52	

The effect of σ-phase formation on the CPT behavior of DSSs is now considered. Aging DSSs at high temperatures generally leads to a decrease in their CPT values [21,22,62]. Figure 22 shows the CPT vs. aging time for S31803 treated at 850 °C in deaerated 1 M NaCl solution. The CPT value of solution-annealed steel is 61 °C and drops considerably with aging time, and reaches a steady value of ~14 °C after aging for 480 min (8 h). For comparison, the impact strength vs.

aging time of S31803 is also plotted in this figure. The impact strength of solution-annealed specimen drops sharply with increasing aging time since the amount of -phase increases with time. For lean UNS S32101 DSS contains very low Mo content, its pitting potential (at 30 °C) and CPT in 1 M NaCl solution also decrease with aging time (Figure 23) [35]. These can be attributed to the formation of Cr_2N precipitates at the / and / grain boundaries during an earlier stage of aging at 700 °C as mentioned previously. Pits nucleate mainly at the Cr-depletion region around the Cr_2N precipitates, and then grow into ferritic grains (Figure 24).

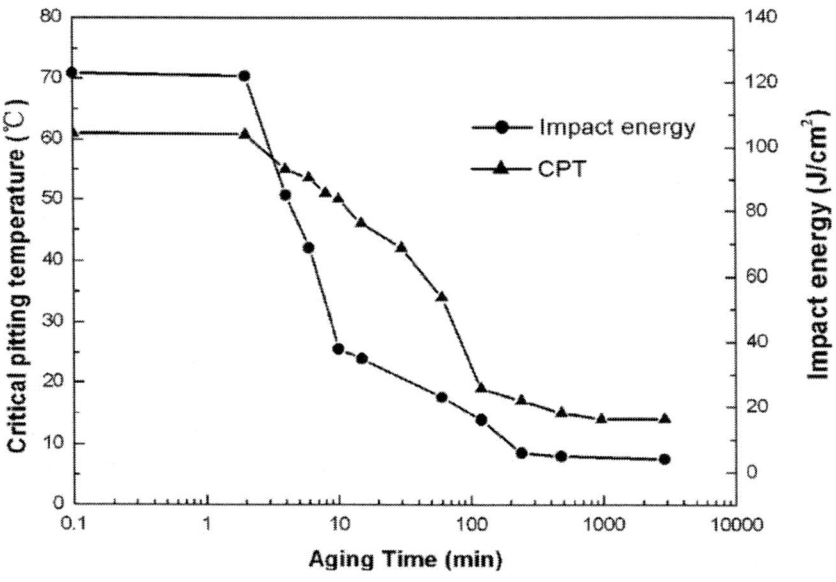

Figure 22: Critical pitting temperature and impact energy *vs.* aging time for UNS S31803, reprinted with permission from [62].

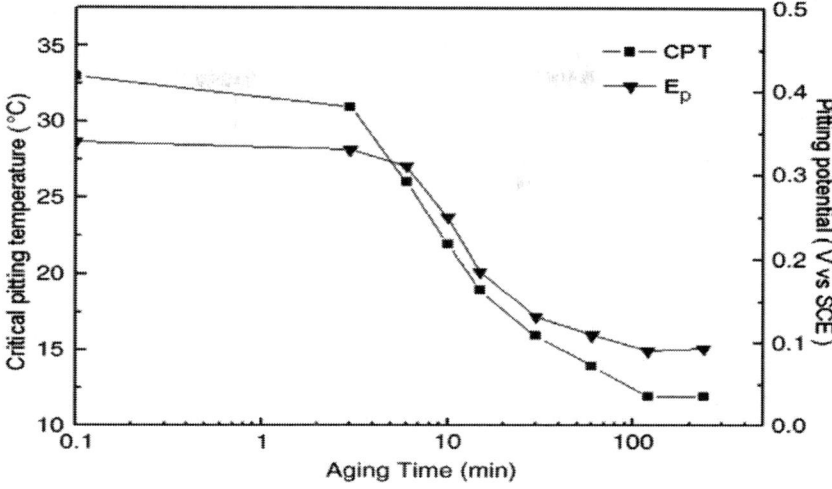

Figure 23: CPT and pitting potential (E_p) of UNS S32101 in 1 M NaCl solution vs. aging time, reprinted with permission from [35].

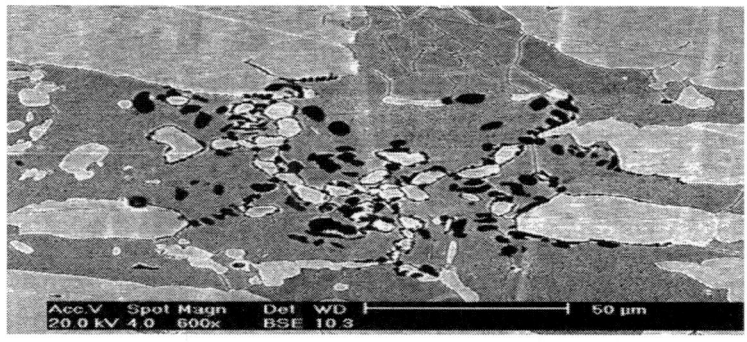

Figure 24: SEM image showing formation of pits in UNS S32101 steel aged at 700 °C for 240 min upon exposure to 1 M NaCl solution. Light phase is austenite and dark phase is ferrite, reprinted with permission from [35].

It is worth noting that spinodal decomposed phases in DSSs also impair their pitting resistance. Figure 25a,b show anodic polarization curves of the as-received and aged DSS 2205 specimens in 1 M HCl solution. The Ecorr of the as-received steel is located at −0.39 V (SCE).

By increasing potential from the Ecorr anodically, passivation occurs and extends from about −0.2 V to 0.85 V (SCE) with a passive current density of 6×10^{-6} A/cm^2. Beyond 0.85V (SCE), pitting occurs due to a dramatic increase in the current density. The passive range and pitting potential of aged samples at 365 °C for different periods and the sample at 400 °C for 500 h remains almost unchanged. However, the passivation behavior is largely disturbed by aging at 365 °C for 5000 h as well as 400 °C for 500 h and 50,000 h. The passive current density of sample aged at 400 °C for 500 h increases to ~2×10^{-5} A/cm^2 (Figure 25b). The pitting potential drops markedly to −0.16 V (SCE) after prolonged aging at 400 °C for 5000 h. This is accompanied by a marked reduction in the passivation range. This is caused by the formation of '-phase, leading to serious pitting attack in the ferrite phase (Figure 26a). Ornek et al. [37] reported that spinodal decomposition is benefical for the corrosion resistance of DSS 2205 aged at 475 °C for up to 10 h because of a better passivation behavior. The CPT value in 3 wt% NaCl solution for the sample aged for 5 h is 50 °C compared with annealed steel of 40 °C. Figure 26b shows a lesser corrosion attack in the -phase of this aged sample after CPT testing. Prolonged aging for 255 h causes a sharp drop in the CPT value to 30 °C owing to the formation of elemental depletion zone [37].

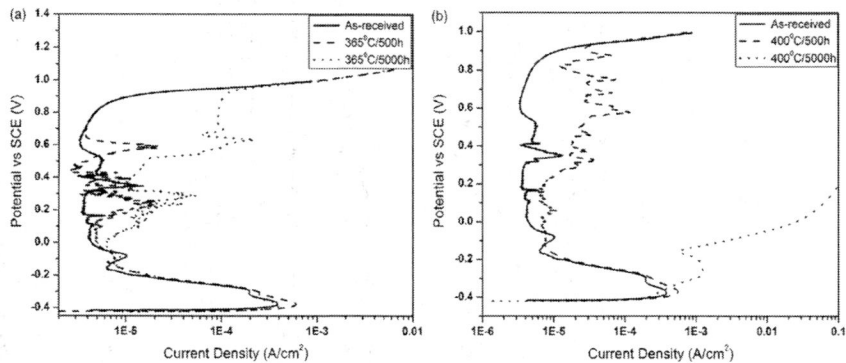

Figure 25: Anodic polarization curves of DSS2205 in 1 M HCl solution aged at (a) 365 °C and (b) 400 °C for different times, reprinted with permission from [39].

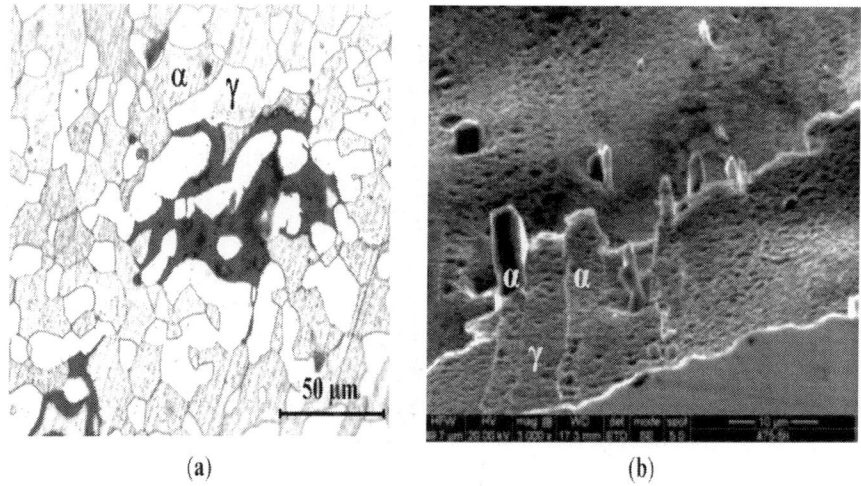

(a)   (b)

Figure 26: (a) Optical micrograph of DSS 2205 aged at 400 °C for 5000 h after polarization testing in 1 M HCl solution, reprinted with permission from [39]. Copyright 2010 Elsevier; (b) SEM image of DSS aged at 475 °C for 5 h after CPT testing, reprinted with permission from [37].

Stress Corrosion Cracking

DSSs generally exhibit higher resistance to SCC in chloride containing solutions than austenitic stainless steels. Therefore, DSSs can replace austenitic grades in the chemical process industries in applications with a high risk of SCC. The resistance of DSSs to SCC is markedly reduced by high temperature, low pH and high applied stress [4]. Constant deformation U-bend, constant load dead-weight, slow strain rate test (SSRT) and fracture mechanics techniques are commonly used for testing SCC susceptibility in aggressive solutions [5,67,68,69,70,71,72,73,74]. In particular, SSRT method (ASTM 129) is performed under tensile loading at low strain rates (e.g., 10^{-6}–10^{-7}/s) [75]. At a critical strain rate, the samples display a ductility minimum, showing the occurrence of brittle failure. The emergence of dislocations at the surface due to their motion along the slip planes disrupts the passive film, resulting in anodic dissolution of metal at highly localized areas. Generally, solution compositions affect the crack initiation process in DSSs. The -phase is more susceptible to SCC attack in hot chloride solutions

[70], while the -phase is selectively attacked in hot alkaline sulfide solutions [71,73]. Besides, microstructural anisotropy of hot rolled DSSs along the rolling longitudinal and transverse longitudinal directions influences the nucleation and growth of fine cracks. This anisotropy affects the SCC susceptibility by favoring crack initiation in the -phase and at phase interfaces [73].

Tsai and Chen reported that annealed DSS 2205 is immune to SCC in a 26 wt% NaCl solution at open-circuit potential (OCP) from 25 to 90 °C, but susceptible to SCC at high anodic potentials [69]. The pitting potential of DSS 2205 in this solution at 90 °C is −160 mV (SCE) (Figure 27a). By applying potentials more noble than this value, the tensile elongation of annealed DSS 2205 reduces markedly (Figure 27b). Thus corrosion pits assist fine crack nucleation, causing selective dissolution of the -phase. In another study, [70], it was found that metal cations affect the activities of Cl⁻ and H⁺ ions for initiating SCC with the ability in the following trend: $Mg^{2+} > Ca^{2+} > Fe^{2+} > Na^+ > Li^+$ [67,68].

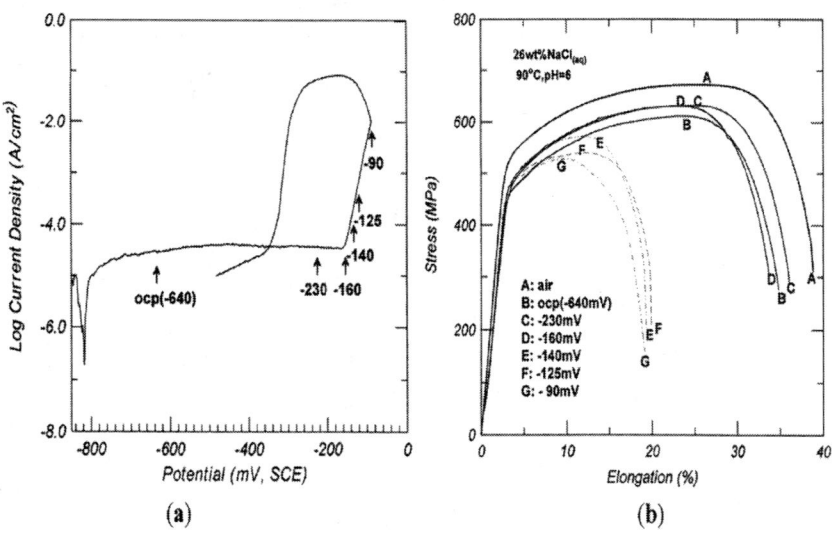

Figure 27: (a) Cyclic polarization curve of annealed DSS 2205 in 26 wt% NaCl solution at 90 °C; (b) Effect of applied potentials on SCC susceptibility of DSS 2205 in 26 wt% NaCl solution at 90 °C at 4.1 × 10^{-6}/s, reprinted with permission from [69].

Recently, Singh-Raman and Siew studied SCC behavior of superduplex SAF 2507 and its suppression in a 30 wt% $MgCl_2$ solution at 180 °C using SSRT method under open-circuit potential condition [72]. This steel experienced intergranular cracking at a strain rate of 3.7 × 10^{-7}/s. To inhibit SCC, $NaNO_2$ inhibitor of different concentrations was added to $MgCl_2$ solution. They reported that the additions of low concentrations of nitrite, *i.e.*, 1400 and 2800 ppm were very effective to suppress SCC due to the NO_2^- ions inhibit pitting of SAF2507 in the $MgCl_2$ solution (Figure 28). Thus pitting assisting crack nucleation was suppressed accordingly. However, addition of 5600 ppm NO_2^- led to the SCC susceptibility, possibly resulting from instability of the passive film at a crack tip.

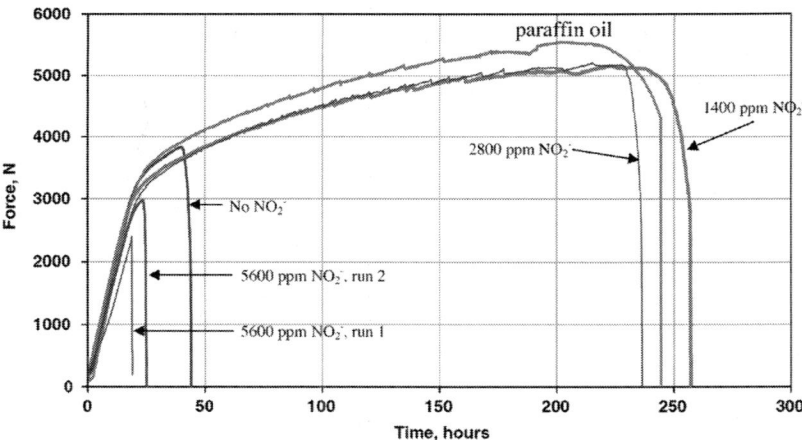

Figure 28: Force *vs.* time slow strain rate test (SSRT) curves for SAF 2507 in 30 wt% $MgCl_2$ solution at 180 °C at a strain rate of 3.7 × 10^{-7}/s with different NO_2^- concentrations, reprinted with permission from [72].

The -phase formed in DSSs is detrimental to their resistance against SCC. Figure 29 shows the SSRT results for aged UNS S31803 in boiling 30 wt% $MgCl_2$ solution at 117 °C (pH = 5). The steel specimens aged at 675 °C for 10 h and at 900 °C for 4 h are particular vulnerable to SCC as revealed by very low tensile elongation of ~5%. More recently, Saithala *et al.* [74] annealed UNS S32750 superduplex at 1080 °C, followed by aging at 850 °C from 45 s up to 12 min. They reported that low levels of -phase (less than 2 vol%) have little effect on the mechanical properties of SDSS exposed to a simulated oil field brine containing carbon dioxide/hydrogen sulfide. At higher levels of -phase (>2 vol%), this steel suffers severe loss of ductility during slow rate tensile straining. The SCC mechanism involves failure of -phase as a result of brittle fracture followed by pitting-assisted anodic dissolution of the - and -phases. Ubhi *et al.* [76] used EBSD for identifying - and -phases in aged S32760, and then performed SSRT of aged steel in chloride solutions at 130 °C. They reported that the - and -phases are detrimental to the resistance of S32760 against SCC. Preferential attack occurred at the - and - boundaries due to the formation of -phase in these regions.

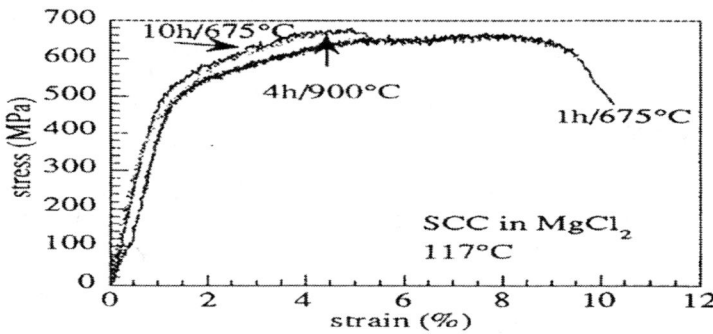

Figure 29: SSRT curves for aged UNS 31803 steel specimens in boiling 30 wt% MgCl$_2$ solution at 117 °C, reprinted with permission from [18].

WELDMENT FAILURE

DSSs find widespread applications in industrial sectors and welding is an important process for fabricating stainless steel structures for engineering applications. DSSs generally have good weldability, but the melting and solidification processes associated with welding destroy favorable duplex microstructure. In addition, welding often causes many defects in the fusion zone (FZ) and heat- affected zone (HAZ) of steel structures [77]. The heat generated during welding process is transferred from the FZ to HAZ. Therefore, FZ and HAZ are the weak points; hence failures of welded components often occur in these regions. The ferrite–austenite phase ratio of DSSs must be maintained close to 50:50 for achieving desired mechanical properties and corrosion resistance. Welding processes tend to upset this phase ratio balance because of slow/fast cooling involved in thermal cycles. The heat input during welding controls the cooling rate and the extent of -ferrite → transformation [78]. During welding, dissolution of primary austenite occurs initially, followed by grain growth in the -ferrite and eventually reformation of austenite during cooling. Austenite tends to nucleate at the grain boundaries, but it can also precipitate in the grain interiors at slow cooling rates [79].

In general, a high heat input or slow cooling rate favors reformation of austenite during cooling. However, intermetallic compounds can form easily in the FZ and HAZ under a slow cooling rate, especially for highly alloyed superduplex and hyperduplex grades [16,80,81]. Consequently, the toughness and corrosion resistance of welded DSSs decrease significantly. On the contrary, a low heat input or fast cooling rate produces an excess of ferrite phase and Cr_2N particles. The solubility of nitrogen in ferrite is quite low, resulting in supersaturation of nitrogen in ferrite and the precipitation of Cr_2N upon rapid cooling from high temperatures [82]. The austenite content in the weld zone can be increased by increasing Ni content of the filler materials and/or the use of nitrogen as a shielding gas, and post-welding heat treatment (PWHT) [83]. The microstructures, mechanical properties and corrosion behaviors of welded DSSs depend greatly on the steel compositions and welding techniques employed. Several techniques are typically used for welding DSSs including tungsten inert gas (TIG) welding, gas metal arc welding, plasma arc welding, friction welding, electron beam welding, laser welding, *etc.* Each technique has its own advantages and limitations [1,2,3,4,16,84,85,86]. Apparently, welding can reduce the corrosion resistance of DSSs by changing the steel composition and microstructure in the FZ. The intermetallic phases and Cr-nitrides induced in weld metals of DSSs can lead to poor corrosion resistance upon exposure in aggressive environments [5,87,88]. In this section, the deleterious effects of these phases on the corrosion behavior of welded DSSs are briefly discussed.

Figure 30a–d show optical micrographs of DSS 2205 welded by the TIG welding process under an argon shielding gas. Three types of austenite are formed in the weld metal during fast solidification: (1) austenite nucleated at the prior ferrite grain boundaries (GBA);

(2) Widmänstten-type austenite (WA) of plate-like feature nucleated from the grain boundaries; and (3) intragranular austenite precipitates in ferritic grains (IGA). The α-phase content of base alloy, weld metal and HAZ determined from image analysis is 49, 77.5 and 75 vol%, respectively. Figure 31a,b is the TEM micrographs showing the formation of rod-like Cr_2N at the α/γ boundaries of fusion zone. These Cr_2N rods grow towards into the ferrite. The SAED pattern and corresponding index diagram are shown in Figure 31c,d, respectively. The formation of Cr_2N depletes Cr content in the α-phase. The high ferrite content together with the formation of Cr_2N rods degrade pitting resistance of the weld metal considerably. In this regard, PWHT can be used to restore a favorable balance of α/γ phase ratio. An optimal PWHT is found to be 1080 °C, reducing the α-content in the weld metal sharply from 77.5 to 53.57 vol%. Consequently, the CPT of the fusion zone immersed in 1 M NaCl solution is 58 °C, i.e., close to that of base alloy with a value of 59 °C. Alternatively, N_2 shielding gas can increase the γ- phase fraction but decrease α-phase and Cr_2N contents in the weld metal of DSSs. Thus the CPT increases while the corrosion rate decreases with increasing N content in the weld metal [89].

Laser beams are coherent and intense, thus capable of attaining fast surface melting followed by rapid solidification. These unique features render the weldments with very fine FZ and HAZ. Recently, Yang et al. [90] studied the microstructure and corrosion behavior of laser welded UNS S31803 steel. The as-welded joint shows poor pitting resistance due to the high volume fraction of ferrite (92%) and the precipitation of Cr_2N in the -phase of fusion zone. CPT measurements in 1 M NaCl solution revealed that the Cr_2N precipitates in FZ are detrimental to the pitting corrosion resistance. The CPT of base metal is 56 °C, but drops sharply to 42 °C in the FZ (Figure 32). During the CPT measurement, a static potential of 0.75 V (SCE) is applied to the specimen while the electrolyte temperature is increased continuously at 1 °C/min. CPT is taken as the current density reaches 100 $\mu A/cm^2$. By examining CPT specimens using optical microscopy and SEM, pits are found to locate preferentially in ferritic grains of weld metal (Figure 33). PWHT also restores a favorable $\alpha:\gamma$ ratio, thereby improving the pitting corrosion resistance of S31803 weldment accordingly.

The HAZ often experiences thermal cycles with temperatures ranging from ambient up to the melting point in regions close to the weld. Furthermore, the heating and cooling rates may vary markedly with heat input, structural dimension, and position relative to the weld [5]. Therefore, the actual HAZ exhibits a complex mix of microstructures within a small volume next to the FZ. Moreover, because the width of HAZ is narrow, it is rather difficult to analyze the effect of a characteristic microstructure on stress corrosion using the true weldment. In this regard, the Gleeble thermo-mechanical simulator allows the simulation of various microstructures developed in HAZ at designed thermal cycles via resistive heating of metal samples [91]. For example, Liou et al. [92] used a Gleeble thermo-mechanical simulator for the welding HAZ simulation of DSS 2205 containing different N contents. The simulated specimens were subjected to U-bend SCC tests in 40 wt% $CaCl_2$ solution at 100 °C. They reported that the GBA, WA and IGA were formed in the HAZ, and their contents varied with cooling rates and N contents in DSS 2205. The HAZ sample with 0.165N exhibited higher austenite and fewer Cr_2N contents, leading to better SCC resistance (Figure 34a). In addition, pitting corrosion assisted the crack initiation, while the types and amounts of reformed austenite in the HAZ affected the mode of crack propagation. The GBA was found to promote intergranular stress corrosion cracking, but WA and IGA exhibited a beneficial effect on stress corrosion by deviating the crack propagation path (Figure 34b). Recently, Singh Raman demonstrated that threshold stress intensity factor for SCC (K_{ISCC}), determined from fracture mechanics testing, is an important parameter in design and prediction of life of welded components [93]. K_{ISCC} is a stress intensity limit below which the crack cannot propagate in a corrosive solution. Determination of K_{ISCC} of narrow HAZ of real welds is difficult with conventional techniques. They recommended circumferential notch tensile technique instead. Alternatively, Gleeble simulator can be used to prepare HAZ samples for the study of crack extension in weldments of DSSs.

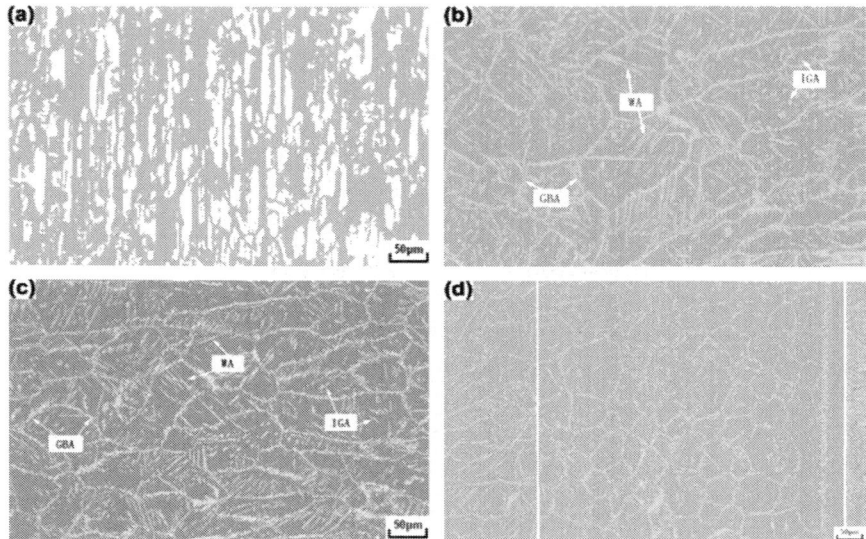

Figure 30: Optical micrographs of (a) as-received DSS 2205; (b) weld metal near fusion line; (c) center of weld metal and (d) heat- affected zone (HAZ) (located between two white lines), reprinted with permission from [87].

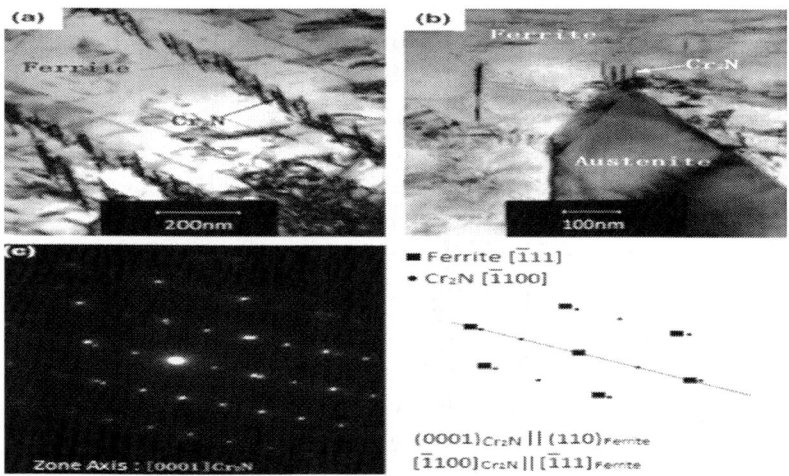

Figure 31: TEM micrographs showing precipitation of Cr_2N rods at (a) fusion zone and (b) α/γ boundaries; (c) SAED pattern and (d) index diagram, reprinted with permission from [87].

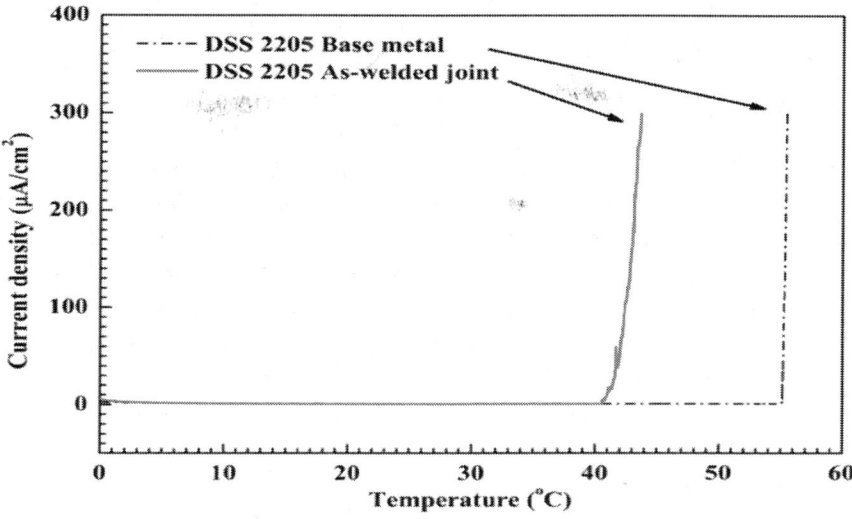

Figure 32: CPT curves of DSS 2205 base metal and laser welded fusion zone, reprinted with permission from [90].

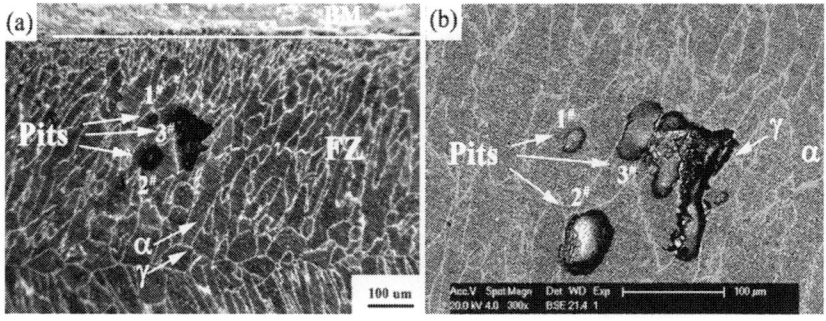

Figure 33: (a) Optical and (b) SEM images of fusion zone of laser welded DSS2205 steel. Austenite is light and ferrite is dark in (a). Pit 1# is located in ferritic grain and pit 2# at the / boundary; large pit 3# spanned across / domains with severely attacked -grains (BM: base metal, FZ: fusion zone), reprinted with permission from [90].

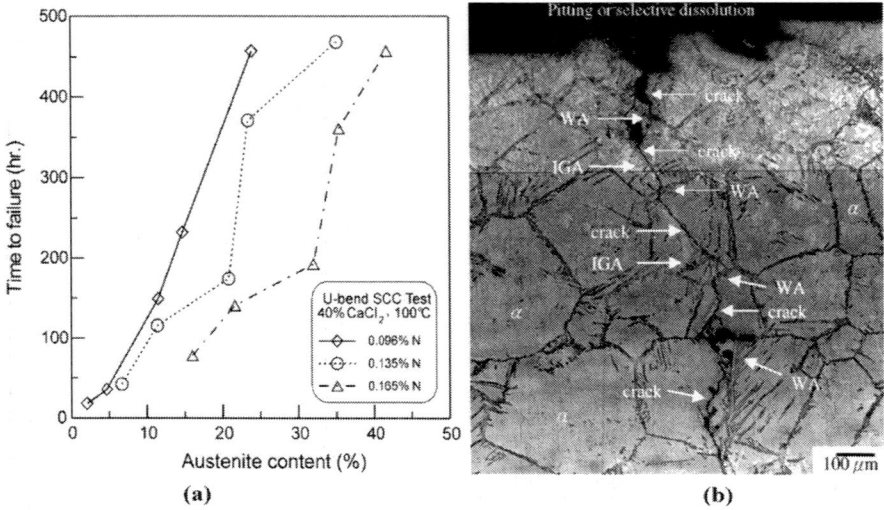

(a)   (b)

Figure 34: (a) Effect of austenite content on time-to-failure of simulated HAZs with various N contents in 40 wt% CaCl$_2$ solution at 100 °C; (b) Cross-sectional optical micrograph showing crack propagation path in the HAZ of DSS 2205 with 0.165N under a simulated cooling rate of ~60 °C/s after U-bend SCC testing, reprinted with permission from [92].

For welded DSSs, cracking and failure often initiated in fusion zone due to the presence of internal defects, chemical inhomogeneity and microstructural modification. Young *et al.* [94] reported that laser-welded DSS 2205 is susceptible to embrittlement in gaseous hydrogen. The hydrogen embrittlement susceptibility was correlated with the microstructures of fusion zone. The susceptibility decreased with increasing austenite content in the weld metal [94]. As mentioned above, sigma phase and chromium nitrides can be induced in the weld metals of DSSs [5,80,81,87,88]. These microstructural features deteriorate the corrosion resistance of weldments, resulting in IGC and pitting corrosion. Furthermore, pitting can assist initiation of fine cracks in DSSs under tensile stress, leading to stress corrosion in DSSs [69,72]. It is considered that pitting initiated in the welds can cause environmental cracking susceptibility largely, leading to catastrophic failure of weldments, particularly in marine environments.

CONCLUSIONS

Duplex stainless steels contain Cr, Mo and N alloying elements exhibiting good corrosion resistance in acidic, caustic and marine environments. However, intermetallic phase formed at 700–900 °C and Cr-rich '-precipitates formed at 350–550 °C are detrimental to the corrosion resistance of DSSs, especially for highly alloyed steels. The formation of these phases depletes Cr or Cr/Mo content in the matrix adjacent to the precipitates, leading to IGC, pitting corrosion and SCC. Such phases are induced in DSSs during the fabrication, improper heat treatment and welding process. Therefore, care must be taken in the alloy design of modern DSSs to ensure optimal loading levels of alloying elements with a stable duplex structure. A balance between the chemical composition and corrosion resistance in DSSs must be maintained to achieve structural integrity.

ACKNOWLEDGMENTS

The authors greatly appreciated the fund granted by the Science Technology and Innovation Committee of Shenzen Municipality (China), and Shenzen Research Institute, City University of Hong Kong, Project No. R-IND4401.

AUTHOR CONTRIBUTIONS

Kai Wang Chan wrote phase precipitation and Sie Chin Tjong wrote corrosion and welding contents of the manuscript. Both authors reviewed and edited the final version.

REFERENCES

1. Davis, J.R. *Stainless Steels*; ASM International: Materials Park, OH, USA, 1994.

2. Gun, R.N. *Duplex Stainless Steels: Microstructure, Properties and Applications*; Woodhead Publishing: Cambridge, UK, 1997.

3.   Alvarez-Armas, I.; Degallaix-Moreuil, S. *Duplex Stainless Steels*; Wiley: Hoboken, NJ, USA, 2009.

4.   Solomon, D.S.; Devine, T.M., Jr. Duplex stainless steel—A tale of two phases. In *Duplex Stainless Steels*; Lula, R.A., Ed.; ASM International: Materials Park, OH, USA, 1983; pp. 693–756.

5.   Nilsson, J.O. Overview of super duplex stainless steel. *Mater. Sci. Technol.* 1992, *8*, 685–700.

6.   Lo, K.H.; Shek, C.H.; Lai, J.K.L. Recent developments in stainless steels. *Mater. Sci. Eng. R.* 2009, *65*, 39–104.

7.   Tjong, S.C.; Hoffman, R.W.; Yeager, E.B. Electron and ion spectroscopic studies of the passive film on iron-chromium alloys. *J. Electrochem. Soc.* 1982, *129*, 1662–1668.

8.   Mesquita, T.J.; Chauveau, E.; Mantel, M.; Nogueira, R.P. A XPS study of the Mo effect on passivation behaviors for highly controlled stainless steels in neutral and alkaline conditions. *Appl. Surf. Sci.* 2013, *270*, 90–97.

9.   Weber, L.; Uggowitzer, P.J. Partitioning of chromium and molybdenum in super duplex stainless steels with respect to nitrogen and nickel content. *Mater. Sci. Eng. A* 1998, *242*, 222–229.

10.   Simmons, J.W. Overview: High-nitrogen alloying of stainless steels. *Mater. Sci. Eng. A* 1996, *207*, 159–169.

11.   Wang, J.; Uggowitzer, P.J.; Magdowski, R.; Speidel, M.O. Nickel-free duplex stainless steels. *Scripta Mater.* 1999, *40*, 123–129.

12.   Merello, R.; Botana, F.J.; Botella, J.; Matres, M.V.; Marcos, M. Influence of chemical composition on the pitting corrosion resistance of non-standard low-Ni high Mn-N duplex stainless steels. *Corros. Sci.* 2003, *45*, 909–921.

13.   Toor, I.; Hyn, P.J.; Kwon, H.S. Development of high Mn–N duplex stainless steel for automobile structural components. *Corros. Sci.* 2008, *50*, 404–410.

14.   Li, J.; Xu, Y.; Xiao, X.; Zhao, J.; Jiang, L.; Hu, J. A new resource-saving, high manganese and nitrogen super duplex stainless steel 25Cr-2Ni-3Mo-xMn-N. *Mater. Sci. Eng. A* 2009, *527*, 245–251.

15.   Li, J.; Ma, Z.; Xiao, X.; Zhao, J.; Jiang, L. On the behavior of nitrogen in a low-Ni high Mn super duplex stainless steel. *Mater. Des.* 2011, *32*, 2199–2205.

16. Karlsson, L. Welding duplex stainless steel—A review of current recommendations. *Weld. World* 2012, *56*, 65–77.

17. Göransson, K.; Nyman, M.L.; Holmquist, M.; Gomes, E. Sandvik SAF 2707 HD (UNS S32707): A hyper-duplex stainless steel for severe chloride containing environments. *Rev. Metall.* 2007, *104*, 411–417.

18. Lopez, N.; Cid, M.; Puiggali, M. Influence of -phase precipitation on mechanical properties and corrosion resistance of duplex stainless steels. *Corros. Sci.* 1999, *41*, 1615–1631.

19. Chen, T.H.; Yang, J.R. Effects of solution treatment and continuous cooling on -phase precipitation in a 2205 duplex stainless steel. *Mater. Sci. Eng. A* 2001, *311*, 28–40.

20. Adhe, K.N.; Kain, V.; Madangopal, K.; Gadiyar, H.S. Influence of sigma phase formation on the localized corrosion behavior of a duplex stainless steel. *J. Mater. Eng. Perform.* 1996, *5*, 500–506.

21. Park, C.J.; Rao, V.S.; Kwon, H.S. Effect of sigma phase on the initiation and propagation of pitting corrosion of duplex stainless steel. *Corrosion* 2005, *61*, 76–83.

22. Ebrahimi, N.; Momeni, M.; Moayed, M.H.; Davoodi, A. Correlation between critical pitting temperature and degree of sensitization on alloy 2205 duplex stainless steel. *Corros. Sci.* 2011, *53*, 637–644.

23. Hong, J.; Han, D.; Tan, H.; Li, J.; Jiang, Y. Evaluation of aged duplex stainless steel UNS S32750 susceptibility to intergranular corrosion by optimized double loop electrochemical potentiokinetic reactivation method. *Corros. Sci.* 2013, *68*, 249–255.

24. Tavares, S.S.M.; Terra, V.F. Corrosion resistance evaluation of the UNS S31803 duplex stainless steels aged at low temperatures (350 to 550 °C) using DLEPR tests. *J. Mater. Sci.* 2005, *40*, 4025–4082.

25. Tjong, S.C.; Lau, K.C. Abrasion resistance of stainless steel composites reinforced with hard TiB_2 particles. *Compos. Sci. Technol.* 2000, *60*, 1141–1146.

26. Lo, K.H.; Kwok, C.T.; Chan, W.K.; Zeng, D. Corrosion resistance of duplex stainless steel subjected to long-term annealing in the spinodal decomposition temperature range. *Corros. Sci.* 2012, *55*, 267–271.

27. Chen, T.H.; Weng, K.L.; Yang, J.R. The effect of high-temperature exposure on the microstructural stability and toughness property in a 2205 duplex stainless steel. *Mater. Sci. Eng. A* 2002, *338*, 259–270.

28. Calliari, I.; Brunelli, K.; Dabala, M.; Ramous, E. Measuring secondary phases in duplex stainless steels. *JOM* 2009, *61*, 80–83.

29. Sieurin, H.; Sandstrom, R. Sigma phase precipitation in duplex stainless steel 2205. *Mater. Sci. Eng. A* 2007, *444*, 271–276.

30. Fargas, G.; Anglada, M. Effect of the annealing temperature on the mechanical properties, formability and corrosion resistance of hot-rolled duplex stainless steel. *J. Mater. Proc. Technol.* 2009, *209*, 1770–1982.

31. Michalska, J.; Sozanska, M. Qualitative and quantitative analysis of and phases in 2205 duplex stainless steel. *Mater. Charact.* 2006, *56*, 355–362.

32. Michalska, J.; Chmiela, B. Phase analysis in duplex stainless steels: Comparison of EBSD and quantitative metallographic methods. *IOP Conf. Series Mater. Sci. Eng.* 2014, *55*.

33. Tan, H.; Jiang, Y.; Deng, B.; Xu, J.; Li, J. Effect of annealing temperature on the pitting resistance of super duplex stainless steel UNS 32750. *Mater. Charact.* 2009, *60*, 1049–1054.

34. Zhang, W.; Jiang, L.; Hu, J.; Song, H. Effect of ageing on precipitation and impact energy of 2101 economical duplex stainless steel. *Mater. Charact.* 2009, *60*, 50–55.

35. Zhang, L.; Jiang, Y.; Deng, B.; Zhang, W.; Xu, J.; Li, J. Effect of aging on the corrosion resistance of 2101 lean duplex stainless steel. *Mater. Charact.* 2009, *60*, 1522–1528.

36. Deng, B.; Jiang, Y.; Xu, J.; Sun, T.; Gao, J.; Zhang, L.; Zhang, W.; Li, J. Application of the modified electrochemical potentiodynamic reactivation method to detect susceptibility to intergranular corrosion of a newly developed lean duplex stainless steel LDX2101. *Corros. Sci.* 2010, *52*, 969–977.

37. Ornek, C.; Engelberg, D.L.; Lyon, S.B.; Ladwein, T.L. Effect of "475 °C embrittlement" on the corrosion behavior of grade 2205 stainless steel investigated using local probing technique. *Corros. Manag. Mag.* 2013, *115*, 9–11.

38. Weng, K.L.; Chen, H.R.; Yang, J.R. The low-temperature aging embrittlement in a 2205 duplex stainless steel. *Mater. Sci. Eng.* 2004, *379*, 119–132.

39. Chandra, K.; Singhal, R.; Kain, V.; Raja, V.S. Low temperature embrittlement of duplex stainless steel: Correlation between mechanical and electrochemical behavior. *Mater. Sci. Eng.* 2010, *527*, 3904–3912.

40. Della Rovere, C.A.; Santos, F.S.; Silva, R.; Souza, C.A.; Kuri, S.E. Influence of long-term low-temperature aging on the microhardness and corrosion properties of duplex stainless steel. *Corros. Sci.* 2013, *68*, 84–90.

41. Lo, K.H.; Lai, J.K. Microstructural characterization and change in ac magnetic susceptibility of duplex stainless steel during spinodal decomposition. *J. Nucl. Mater.* 2010, *401*, 143–148.

42. Hyde, J.M.; Miller, M.K.; Hetherington, M.G.; Cerezo, A.; Smith, G.D.; Elliot, C.M. Spinodal decomposition in Fe-Cr alloys: Experimental study at the atomic level and comparison with computer models-II. *Acta Metall. Mater.* 1991, *43*, 3403–3413.

43. Hetherington, M.G.; Hyde, J.M.; Miller, M.K.; Smith, G.D. Measurement of the amplitude of a spinodal. *Surf. Sci.* 1991, *246*, 304–314.

44. Brown, J.E.; Smith, G.D. Atom probe study of spinodal processes in duplex stainless steels and in single- and dual phase Fe-Cr-Ni alloys. *Surf. Sci.* 1991, *246*, 285–291.

45. Miller, M.K. The development of atom probe field-ion microscopy. *Mater. Charact.* 2000, *44*, 11–27.

46. Danoix, F.; Auger, P. Atom probe studies of the Fe-Cr system and stainless steels aged at intermediate temperature: A review. *Mater. Charact.* 2000, *44*, 177–201.

47. Bhattacharya, A.; Singh, P.M. Electrochemical behavior of duplex stainless steels in caustic environment. *Corros. Sci.* 2011, *53*, 71–81.

48. Thierry, D.; Persson, D.; Leygraf, C.; Delichere, D.; Joiret, S.; Pallotta, D.; Golf, A. *In-situ* Raman spectroscopy combined with X-ray photoelectron spectroscopy and nuclear microanalysis of anodic corrosion film formation on Fe-Cr single crystals. *J. Electrochem. Soc.* 1988, *135*, 305–310.

49. *Standard Test Methods for Detecting Detrimental Intermetallic Phase in Duplex Austenitic/Ferritic Stainless Steels*; ASTM A923–03. American Society for Testing and Materials: West Conshohocken, PA, USA, 2003.

50. *Standard Test Method for Electrochemical Reactivation (EPR) for Detecting Sensitization of AISI Type 304 and 304L Stainless Steels*; American Society for Testing and Materials: West Conshohocken, PA, USA, 1994.

51. Clarke, W.L.; Romero, V.M.; Danko, J.C. Detection of sensitization in stainless steels using electrochemical techniques. *Corrosion* 1977, *77*, 130–133.

52. Majidi, A.P.; Streicher, M.A. The double loop reactivation method for detecting sensitization in AISI-304 stainless steel. *Corrosion* 1984, *40*, 584–593.

53. *Corrosion of Metals and Alloys—Electrochemical Potentiokinetic Reactivation Measurement Using the Double Loop Method (Based on Cihal's Method)*; ISO 12732:2006. International Organization for Standardization: Geneva, Switzerland, 2009.

54. Cihal, V.; Stefec, R. On the development of electrochemical potentiokinetic method. *Electrochim. Acta* 2001, *46*, 3867–3877.

55. Ortiz, N.; Curiel, F.F.; Lopez, V.H.; Ruiz, A. Evaluation of the intergranular corrosion susceptibility of UNS S31803 duplex stainless steel with thermoelectric power measurements. *Corros. Sci.* 2013, *69*, 236–244.

56. Amadou, T.; Braham, C.; Sidhom, H. Double loop electrochemical potentiokinetic reactivation test optimization in checking of duplex stainless steel intergranular corrosion susceptibility. *Metall. Mater. Trans. A* 2004, *35*, 3499–3513.

57. Gong, J.; Jiang, Y.M.; Deng, B.; Xu, J.L.; Hu, J.P.; Li, J. Evaluation of intergranular corrosion susceptibility of UNS S31803 duplex stainless steel with an optimized double loop electrochemical potentiokinetic reactivation method. *Electrochim. Acta* 2010, *55*, 5077–5083.

58. Lo, K.H.; Kwok, C.T.; Chan, W.K. Characterization of duplex stainless steel subjected to long-term annealing in the sigma phase formation temperature range by the DLEPR test. *Corros. Sci.* 2011, *53*, 3697–3703.

59. Fontana, M.G. *Corrosion Engineering*; McGraw-Hill: New York, NY, USA, 1987.

60. Zhang, Z.; Han, D.; Jiang, Y.; Shi, C.; Li, J. Microstructural evolution and pitting resistance of annealed lean duplex stainless steel UNS S32304. *Nucl. Eng. Des.* 2012, *243*, 56–62.

61. Deng, B.; Jiang, Y.; Gong, J.; Gao, J.; Li, J. Critical pitting and repassivation temperatures for duplex stainless steel in chloride solutions. *Electrochim. Acta* 2008, *53*, 5220–5225.

62. Deng, B.; Wang, Z.; Jiang, Y.; Wang, H.; Gao, J.; Li, J. Evaluation of localized corrosion in duplex stainless steel aged at 850 °C with critical pitting temperature measurement. *Electrochim. Acta* 2009, *54*, 2790–2794.

63. Ebrahimi, N.; Momeni, M.; Kosari, A.; Zakeri, M.; Moayed, M.H. A comparative study of critical pitting temperature (CPT) of stainless steels by electrochemical impedance spectroscopy (EIS), potentiodynamic and potentiostatic techniques. *Corros. Sci.* 2012, *59*, 96–102.

64. Domínguez-Aguilar, M.A.; Newman, R.C. Detection of deleterious phases in duplex stainless steel by weak galvanostatic polarization in halide solutions. *Corros. Sci.* 2006, *48*, 2577–2591.

65. Brigham, R.J.; Tozer, E.W. Temperature as a pitting criterion. *Corrosion* 1973, *29*, 33–36.

66. *Standard Test Method for Electrochemical Critical Pitting Temperature Testing of Stainless Steels*; ASTM G150–13. American Society for Testing and Materials: West Conshohocken, PA, USA, 2013.

67. Prosek, T.; Iversen, A.; Taxen, C.; Thierry, D. Low-temperature stress corrosion cracking of stainless steels in the atmosphere in the presence of chloride deposits. *Corrosion* 2009, *65*, 105–117.

68. Kangas, P.; Nicholls, J.M. Chloride induced stress corrosion cracking of duplex stainless steels: Models, test methods and experience. *Mater. Corros.* 1995, *46*, 354–365.

69. Tsai, W.T.; Chen, M.S. Stress corrosion cracking behavior of 2205 duplex stainless steel in concentrated NaCl solution. *Corros. Sci.* 2000, *42*, 545–559.

70.   Tseng, C.M.; Liou, H.Y.; Tsai, W.T. Effect of nitrogen content of the environmentally-assisted cracking susceptibility of duplex stainless steel. *Metall. Mater. Trans. A* 2003, *34*, 95–103.

71.   Bhattacharya, A.; Singh, P.M. Effect of heat treatment on corrosion and stress corrosion cracking of S32205 duplex stainless steel in caustic solutions. *Metall. Mater. Trans. A* 2009, *40*, 1388–1399.

72.   Singh Raman, R.K.; Siew, W.H. Role of nitrite addition in chloride stress corrosion cracking of a super duplex stainless steel. *Corros. Sci.* 2010, *52*, 113–117.

73.   Chasse, K.R.; Singh, P.M. Effect of microstructural anisotropy on stress corrosion cracking of hot rolled duplex stainless steels. *Corros. Eng. Sci. Technol.* 2012, *47*, 170–178.

74.   Saithala, J.R.; Mahajanam, S.; Ubhi, H.S.; Atkinson, J. Environmental-assisted cracking behavior of sigmatized super duplex stainless steel in oilfield production line. *Corrosion* 2013, *69*, 276–285.

75.   *Standard Practice for Slow Strain Rate Testing to Evaluate the Susceptibility of Metallic Materials to Environmentally Assisted Cracking*; ASTM G129–00. American Society for Testing and Materials: West Conshohocken, PA, USA, 2000.

76.   Application of the EBSD technique in corrosion studies of super duplex stainless steels. Available online: http://www.oxford-instruments.com/getmedia/bea996d1-4617-4df2-bc23-2281753bdd9f/application-of-the-EBSD-technique-in-corrosion-studies-of-super-duplex-stainless-steels (accessed on 18 July 2014).

77.   Nowacki, J.; Rybicki, P. The influence of welding heat input on submerged arc welded duplex steel joints imperfections. *J. Mater. Proc. Technol.* 2005, *164–165*, 1082–1088.

78.   Baeslack, W.A, III; Lippold, J.C. Phase transformation behavior in duplex stainless steel weldments. *Met. Constr.* 1988, *20*, 26R–31R.

79.   Seurin, H.; Sandstrom, R. Austenite reformation in the heat-affected zone of duplex stainless steel 2205. *Mater. Sci. Eng. A* 2006, *418*, 250–256.

80. Karlsson, L.; Pak, S.; Ryen, L. Precipitation of intermetallic phases in 22% Cr duplex stainless weld metals. *Weld. J.* 1995, *74*, 28–39.

81. Sato, Y.S.; Kokawa, H. Preferential precipitation of sigma phase in duplex stainless steel weld metal. *Scr. Mater.* 1999, *40*, 659–653.

82. Muthupandi, V.; Bala Srinivasan, P.; Sundaresan, S. Effect of weld chemistry and heat input on the structure and properties of duplex stainless steel welds. *Mater. Sci. Eng. A* 2003, *358*, 9–16.

83. Igual Munoz, A.; Garcia Anton, J.; Guinon, J.L.; Perez Herranz, V. Effect of nitrogen in argon as a shielding gas on TIG welds of duplex stainless steels. *Corrosion* 2005, *61*, 693–705.

84. Urena, A.; Otero, E.; Utrilla, M.V.; Munez, C.J. Weldability of a 2205 duplex stainless steel using plasma arc welding. *J. Mater. Proc. Technol.* 2007, *182*, 624–631.

85. Zambon, A.; Bonollo, F. Rapid solidification in laser welding of stainless steels. *Mater. Sci. Eng. A* 1994, *178*, 203–207.

86. Gooch, T.G. Weldability of duplex ferritic-austenitic stainless steel. In *Duplex Stainless Steels*; Lula, R.A., Ed.; ASM International: Materials Park, OH, USA, 1983; pp. 573–602.

87. Zhang, Z.; Wang, Z.; Jiang, Y.; Tan, H.; Han, D.; Guo, Y. Effect of post-weld heat treatment on microstructure evolution and pitting corrosion behavior of UNS S31803 duplex stainless steel welds. *Corros. Sci.* 2012, *62*, 42–50.

88. Turnbull, A.; Francis, P.E.; Ryan, M.P.; Orkney, L.P.; Griffith, A.J.; Hawkins, B. A novel approach to characterizing corrosion resistance of super duplex stainless steel weld. *Corrosion* 2002, *58*, 1039–1048.

89. Matsunaga, H.; Sato, Y.S.; Kokawa, H.; Kuwana, T. Effect of nitrogen on corrosion of duplex stainless steel weld metal. *Sci. Technol. Weld. Join.* 1998, *3*, 225–232.

90. Yang, Y.; Wang, Z.; Tan, H.; Hong, J.; Jiang, Y.; Jiang, L.; Li, J. Effect of a brief post-weld heat treatment on the microstructure evolution and pitting corrosion of laser beam welded UNS S31803 duplex stainless steel. *Corros. Sci.* 2012, *65*, 472–470.

91. Adonyi, Y. Heat-affected zone characterization by physical simulations. *Weld. J.* 2006, *85*, 42–47.

92.   Liou, H.Y.; Hsieh, R.I.; Tsai, W.T. Microstructure and stress corrosion cracking in simulated heat-affected zone of duplex stainless steels. *Corros. Sci.* 2002, *44*, 2841–2856.

93.   Singh Raman, R.K. Advances in techniques for determination of susceptibility of welds to stress corrosion cracking (K_{ISCC}). In *Weld Cracking in Ferrous Alloys*; Singh Raman, R.K., Ed.; Woodhead Publishing: Cambridge, UK, 2009; pp. 521–533.

94.   Young, M.C.; Tsay, L.W.; Shin, C.S. Hydrogen-enhanced cracking of 2205 duplex stainless steel welds. *Mater. Chem. Phys.* 2005, *91*, 21–27.

Chapter

6

Comparative Study on Corrosion Protection of Reinforcing Steel by Using Amino Alcohol and Lithium Nitrite Inhibitors

Han-Seung Lee[1], Hwa-Sung Ryu[1], Won-Jun Park[2], and Mohamed A. Ismail[1]

[1]Department of Architectural Engineering, Hanyang University, 1271 Sa 3-dong, Sangrok-gu, Ansan 426-791, Korea

[2]Sustainable Building Research Center, Hanyang University, 1271 Sa 3-dong, Sangrok-gu, Ansan 426-791, Korea

ABSTRACT

In this study, the ability of lithium nitrite and amino alcohol inhibitors to provide corrosion protection to reinforcing steel was investigated. Two types of specimens—reinforcing steel and a reinforced concrete prism that were exposed to chloride ion levels resembling the chloride attack environment—were prepared. An autoclave accelerated corrosion test was then conducted. The variables tested included the chloride-ion concentration and molar ratios of anti-corrosion ingredients in a $CaOH_2$-saturated aqueous solution that simulated a cement-pore solution. A concentration of 25% was used for the lithium nitrite inhibitor $LiNO_2$, and an 80% solution of dimethyl ethanolamine $((CH_3)_2NCH_2CH_2OH$, hereinafter DMEA) was used for the amino alcohol inhibitor. The test results indicated that the lithium nitrite inhibitor displayed anti-corrosion properties at a molar ratio of inhibitor of ≥ 0.6; the amino alcohol inhibitor also displayed anti-corrosion properties at molar ratios of inhibitor greater than approximately 0.3.

INTRODUCTION

Corrosion of reinforcing steel in concrete arising from the penetration of chlorides, which can result from the exposure of concrete to marine environments or ocean sand, represents one of the major factors that cause deterioration of reinforced concrete structures [1]. Reinforcing steel in reinforced concrete structures is protected by an alkali-oxide film formed during the cement hydration process [2]. However, if the chloride concentration at the reinforcing steel surface reaches a critical value, the film on the passive-state metal is destroyed, which triggers corrosion of the reinforcing steel. Chlorides in concrete come from seawater, the salinity of airborne chlorides, or de-icing salts. The infiltration of chlorides inside the concrete is controlled by the concrete's characteristics, such as the unit amount of cement, the ratio of water to binding material, and the type of binding material used [3]. As a method of inhibiting reinforcing steel corrosion inside concrete from chlorides, the use of corrosion inhibitors during concrete preparation is becoming common practice. However, the anti-corrosion effects and proper amount of corrosion inhibitors used to inhibit reinforcing steel corrosion in reinforced concrete structures exposed to a chloride environment need to be examined. Quantitative data on the anti-corrosion effects, according to the type of corrosion inhibitor used, are difficult to obtain. Moreover, since corrosion inhibitors experience a dramatic decline in performance at high temperatures and pressures, depending on the type, new methods for evaluating the anti-corrosion properties of these inhibitors and further studies are desperately needed [4].

Reinforcing steel corrosion by chlorides acts as a major deterioration factor at the interface between reinforcing steel and concrete inside the concrete structure. The reinforcing steel corrosion is initiated once the chloride concentration exceeds a certain critical value, and it is a well-known fact that chloride ions are involved in the process of reinforcing steel corrosion [5]. There have been many studies on the critical chloride concentration. A precedent study indicated that in case the concentration ratio of $[Cl^-:OH^-]$ exceeds 0.6, damage to the film on the passive-state metal occurs, initiating local corrosion [6].

Generally, corrosion inhibitors are classified as anodic or mixed reactants, depending on their reaction mechanism. Anodic corrosion inhibitors, which are mostly nitrite-based, display corrosion protection through a partial interface process. Oxidation of ferric ions forms a ferric-oxide film around the reinforcing steel. Nitrite-based inhibitors are considered the most effective products on the market. Concerns are with their toxicity, solubility, and possible increase of corrosion rate in the case of low dosage or in the presence of cracks, and also the relatively high costs of this type of additive [7,8,9,10]. Thus, the use of amino alcohol-based mixed corrosion inhibitors has increased recently, but studies related to these types of inhibitors are lacking.

A pilot study was initiated to evaluate the anti-corrosion properties of these two corrosion inhibitors, qualitatively. Tests were conducted on lithium nitrite and amino alcohol inhibitors by means of electrochemical techniques. An autoclave accelerated corrosion test that simulates accelerated reinforcing steel corrosion inside concrete was conducted [11,12,13]. The anti-corrosion effects were evaluated by measuring the speed of the reinforcing steel corrosion using a Tafel plot. The investigated variables included the chloride ion concentration and the molar ratio of inhibitor to chloride in a $CaOH_2$-saturated aqueous solution that simulated a cement-pore solution. The study was intended to confirm the anti-corrosion properties of these corrosion inhibitors in relation to the chloride-ion content inside concrete [14]. The electrochemical evaluation methods used to access reinforcing steel corrosion are receiving much attention as they represent non-destructive techniques of relatively high accuracy and offer the possibility to measure small amounts of corrosion within a short period of time [15,16]. Among them, the techniques most frequently used are half-cell potential mapping, concrete resistivity, liner polarization resistance, and A/C impedance [17,18].

It is planned after confirming the outcomes of this study to launch a comprehensive program to study the anti-corrosion effects of these inhibitors on different specimens under different environments on a long-term basis.

MECHANISMS OF LITHIUM NITRITE AND AMINO ALCOHOL INHIBITORS

Lithium Nitrite Inhibition Mechanism

Reinforcing steel corrosion in a reinforced concrete structure is prevented by the formation of a film on passive state metals on the reinforcing steel surface by the high alkalinity of the concrete. This film on passive state metals (Fe_2O_3) is stable, with strong bonding in an alkaline environment, but it becomes unstable and dissolves in the presence of chloride ions. Accordingly, the anti-corrosion mechanism of the lithium nitrite inhibitor commonly used in reinforced concrete structures involves nitrite ions (NO_2^-) from nitrite reacting with ferrous ions (Fe^{2+}), which inhibits the movement of Fe^{2+} from the anode, whereby Fe_2O_3 is deposited on the iron surface to form a film on passive state metals. As a result, the corrosion reaction is inhibited. The reaction follows Equation (1) and the mechanism is shown in Figure 1 [19].

$$2Fe^{2+} + 2OH^- + 2NO_2^- \rightarrow 2NO_2 \uparrow + Fe_2O_3 + H_2O \qquad (1)$$

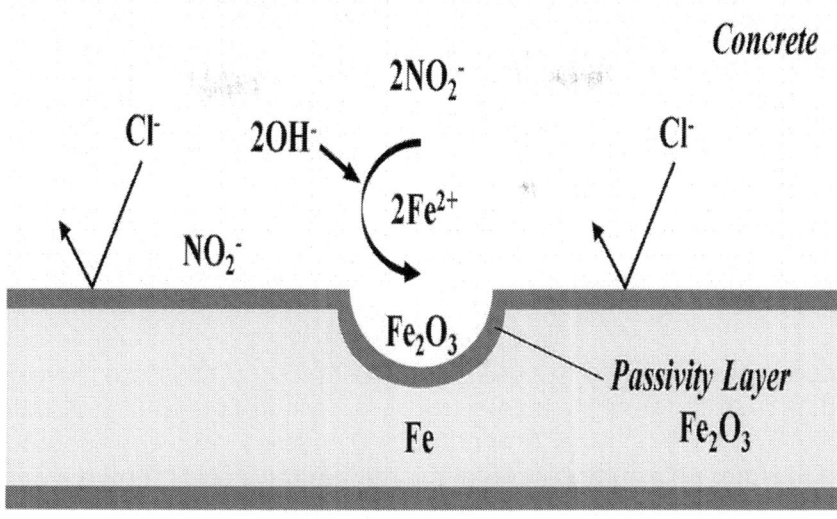

Figure 1: Anti-corrosive mechanism of nitrite-based corrosion inhibitors.

Amino Alcohol Inhibitor Mechanism

Amino alcohol corrosion inhibitors control corrosion by attacking the cathodic activity, blocking sites where oxygen picks up electrons and is reduced to hydroxyl ion. Also, inhibition of corrosion occurs through a mechanism whereby amino alcohols displace chloride ions and form a durable passivating film. In this view, although the amino alcohols adsorb on non-corroding sites, which may seem more cathodic than anodic, they can just as easily be said to adsorb on potentially anodic sites as well [20]. Ormellese *et al.* [21] suggested that organic corrosion inhibitors reduce the ingress of chloride by filling concrete pores and blocking the porosity of concrete by the formation of complex compounds. Thus, the value of chlorides reaching to the steel surface is significantly less so corrosion is inhibited. Several studies of the corrosion inhibition effect of amino alcohols on steel report their performance as a function of concentration and pH in saline solutions [22, 23, 24].

TESTING ANTI-CORROSION CHARACTERISTICS OF CORROSION INHIBITORS

Test Summary

The present study assessed the electrochemical characteristics of corrosion inhibitors in aqueous solution based on the inhibitor amounts added in order to perform a qualitative evaluation of their anti-corrosion effects First, in terms of the electrochemical anti-corrosion properties, Table 1 lists the physical properties of the tested corrosion inhibitors, and the chemical composition of the reinforcing steel is shown in Table 2.

The potentiodynamic polarization curve, known as a Tafel plot, depicts the relationship between potential and corrosion current density. This plot exhibits a linear region, the slope of which is known as the Tafel constants (anodic and cathodic Tafel constants). The intersection of the projection of the linear region of the plot with the open circuit potential (E_{corr}) gives the cathodic or anodic corrosion current (i_{corr}). Once i_{corr} is determined, the following equation, derived from Faraday's law, can be used to calculate the corrosion rate [25]:

$$\text{Corrosion rate } (\mu m/y) = \frac{3.27 \times I_{corr} \times E.W.}{d} \tag{2}$$

The corrosion rate in Equation (2) is expressed in micrometers per year, $\mu m/y$. I_{corr} is the corrosion current density in $\mu A \cdot cm^{-2}$, obtained by dividing i_{corr} with the exposed surface area of the measured specimen. $E.W.$ is the equivalent weight of steel in g, and d is the density of steel in g/cm^3.

The polarization resistance R_p ($\Delta E / \Delta I$), which is the slope of the potential-current curve at E_{corr}, is related to I_{corr} through the following Stern–Geary relationship [26]:

$$I_{corr} \ (\mu A/cm^2) = \frac{[\beta_a \times \beta_c]}{2.3 \ (\beta_a + \beta_c)} R_p$$

(3)

β_a and β_c are the anodic and the cathodic Tafel constants, respectively, expressed in mV/decade of the current. R_p is expressed in $K\Omega \cdot cm^2$. It is seen here that for the determination of I_{corr} in this technique, β_a and β_c are determined from the Tafel plot.

A potentiostat was used to measure the corrosion potential (E_{corr}), corrosion current density (I_{corr}), and corrosion rate (CR). Also, the pore solution was prepared by adding NaCl to saturated calcium hydroxide (solubility of 0.173 g/100 mL at 20 °C). Concentrations of chloride ions (NaCl amount added) were set to 0.6 kg/m³, 1.2 kg/m³, 2.4 kg/m³, and 4.8 kg/m³, with the chloride-ion content used as standard for the prediction of the service life set to 1.2 kg/m³.

The types of corrosion inhibitors used were lithium nitrite ($LiNO_2$) and DMEA (($CH_3)_2NCH_2CH_2OH$). The molar ratios of the anti-corrosion ingredients, based on the ratio of chloride to hydroxide ions, were set to 0.0, 0.3, 0.5, and 1.2 for the experiments. The lithium nitrite inhibitor used was a 25% solution of $LiNO_2$ and the amino alcohol inhibitor used was an 80% concentrated solution. The amounts to be added were calculated and the tests were performed accordingly. Table 3shows the different experimental parameters and their values. With respect to the chloride ion concentrations of 1.2 kg/m³ and 2.4 kg/m³, the chloride ion-dependent molar ratios of inhibitor to chloride were set to 0.0, 0.3, 0.6, and 1.2.

Table 1: Physical properties of inhibitors

Type	Main Component of Inhibitor	Specific Gravity	pH	Viscosity (cps)	Solid Content (%)
Lithium Nitrite inhibitor	LiNO2	1.12	11.5	13	25
Amino Alcohol inhibitor	(CH3)2NCH2CH2OH	1.07	11.9	11	80

Table 2: Chemical composition of reinforcing steel (%)

C	Si	Mn	P	S	Ni	Cr	Mo	Cu	Sn
0.24	0.26	0.95	0.016	0.008	0.03	0.04	0.01	0.02	0.0005

Table 3: Experimental testing parameters conditions

No.	Content of Cl− (kg/m3)	Lithium Nitrite Inhibitor		Amino Alcohol Inhibitor	
		LiNO2		(CH3)2NCH2CH2OH	
		Molar Ratio	Addition	Molar Ratio	Addition
		[Cl−]/[NO2−]	kg/m3	[Cl−]/[OH−]	kg/m3
1	0.0	0.0	0.00	0.0	0.00
2	0.6	0.0	0.00	0.0	0.00
3		0.3	1.08	0.3	0.45
4		0.6	2.15	0.6	0.91
5		1.2	4.30	1.2	1.81
6	1.2	0	0.00	0.0	0.00
7		0.3	2.15	0.3	0.91
8		0.6	4.30	0.6	1.81
9		1.2	8.60	1.2	3.62
10	2.4	0.0	0.00	0.0	0.00
11		0.3	4.30	0.3	1.81
12		0.6	8.60	0.6	3.62
13		1.2	17.21	1.2	7.24
14	4.8	0.0	0.00	0.0	0.00
15		0.3	8.60	0.3	3.62
16		0.6	17.21	0.6	7.24
17		1.2	34.41	1.2	14.48

Tests were carried out first on reinforcing steel specimens in solutions that contain different inhibitor and NaCl concentrations to determine the best molar ratio for the inhibitor before the second stage was carried out. The second stage involved studying the effect of the best inhibitor molar ratio calculated from stage one on the anti-corrosion protection of reinforcing steel in concrete that was subjected to chloride attack.

To suggest an effective measure for reinforcing steel corrosion inhibition inside concrete, the reinforcing steel corrosion conditions were examined after corrosion acceleration of the reinforced concrete had taken place. The corrosion acceleration of the reinforcing steel in concrete was done using an autoclave. The autoclave method is an accelerated corrosion method that conforms to Korean standard KS F 2599-1 [11].

For a reinforced concrete specimen, the corrosion potentials were measured by using the half-cell potential technique, and the corrosion-area ratio was calculated by confirming the reinforcing steel corrosion conditions after the application of the accelerated corrosion method for reinforced concrete. The amount of water-soluble chloride, which directly affects reinforcing steel corrosion, was measured using a potentiometric titration apparatus, in accordance with ASTM C 1218 [27]. Then, the results were analyzed. Table 4 shows the composition and the physical properties of the cement and Table 5 shows the physical properties of the aggregates. The mix proportion of the concrete is indicated in Table 6. The tests were conducted using coarse aggregate with a nominal maximum size of 25 mm, a water to cement ratio of 0.60, and a unit cement amount of 300 kg/m^3. The size of the test specimens was 40 × 40 × 160 mm^3.

Table 4: Chemical composition and physical properties of cement LOI: Loss on ignition

Chemical Composition (%)							Specific Surface (cm2/g)
SiO2	Al2O3	Fe2O3	CaO	MgO	SO3	LOI	
21.95	6.59	2.81	60.12	3.32	2.11	2.58	3.14

Table 5: Physical properties of aggregates

Type	Density (kg/m3)	Absorption (%)	Fineness Modulus
Fine aggregate	2.58	1.34	2.57
Coarse aggregate	2.70	1.4	6.83

Table 6: Mix proportion of concrete W/C: water–cement ratio; S/a: fine aggregate percentage

W/C (%)	Air (%)	S/a (%)	Weight Mixing (kg/m3)				
			Water (kg/m3)	Cement (kg/m3)	Fine Aggregates (kg/m3)	Coarse Aggregates (kg/m3)	Admixture (C × %)
60	4.5	43	186	300	836	1033	0.5

Electrochemical Testing of Anti-Corrosion Properties

To measure the corrosion status of reinforcing steel in NaCl solution, a circular reinforcing steel section with a diameter of 10 mm was cut to a length of 10 mm and a wire was welded on one side to provide an electrical contact. After that, the entire specimen was insulated with a silicone coating except for the areas needed for the measurement of the electrochemical characteristics. The prepared specimen was soaked in $CaOH_2$ solution for 10 min in order to reach a stable state before testing. The electrochemical properties, such as the corrosion potential and corrosion rate, were then measured. Figure 2 shows photos of the reinforcing steel specimen before and after the accelerated corrosion procedure and a schematic diagram of the test rig used in the experiments. Figure 3 shows the details of the reinforced concrete specimen.

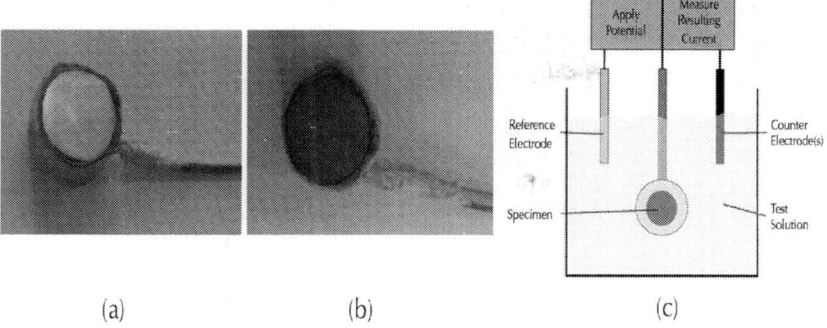

(a)   (b)   (c)

Figure 2: Reinforcing steel specimen and schematic diagram of the test rig. (a) Specimen (Before); (b) Specimen (After); (c) Schematic diagram of the potentiostat.

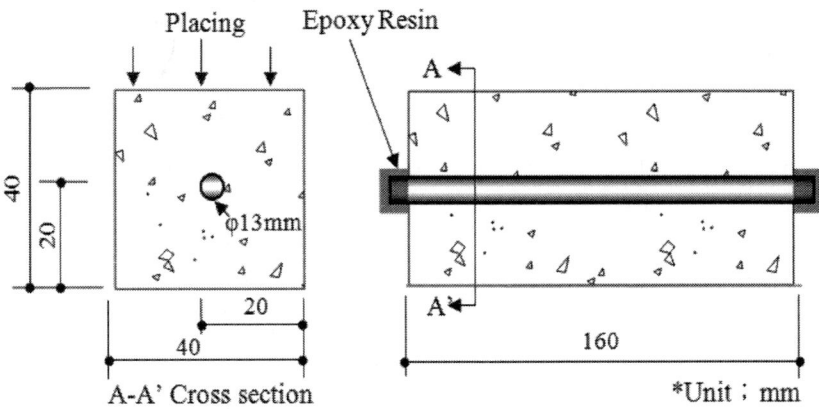

Figure 3: Details of test specimen.

Interpretation of Tafel Plot Curves

Qualitative determination of the level of corrosion protection of the reinforcing steel can be done by calculating the corrosion current density as in Equation (2). The corrosion rate then can be calculated as in Equation (1). In the experiments, 10 mV/min was used as the potential sweep rate for the corrosion current density measurements and the potential-difference range was set to 0–1000 mV.

Experiments on Reinforcing Steel Anti-Corrosion Properties by the Accelerated-Corrosion Method

As shown in Figure 4, the corrosion of a reinforced concrete specimen was accelerated by the autoclave accelerated method, in accordance with autoclave accelerated corrosion test [11]. Figure 5 shows the procedure for measuring the reinforcing steel anti-corrosion properties with: "Figure 5a" measuring the corrosion potential; "Figure 5b" the autoclave-accelerated corrosion test, and then the determination of the soluble chloride content using the potentiometric titration apparatus.

After the completion of the accelerated corrosion tests, the corrosion potentials were measured by the half-cell potential technique, as shown in Figure 5, and the corrosion/area ratio was estimated by scanning the corrosion shape with a transparent sheet and using automated area-measurement software.

For measuring the water-soluble chloride content, which has a direct impact on reinforcing steel corrosion, a potentiometric titration apparatus was used in accordance with the ASTM C 1218 guidelines [27]. For the measurements of the water-soluble chloride content after accelerated corrosion, the test specimen was split using a universal testing machine, and a 10 g powder sample at a distance of 10 mm from the surface of the concrete was collected, again in accordance with ASTM C 1218 [27]. The 10-g powder sample was well mixed in a beaker with 50 mL of ion-exchange solution. Next, approximately 3 mL of hydrogen peroxide (H_2O_2) was added to eliminate any effects of sulfides. To prevent any changes in the chloride concentration originating from evaporation of the sample filtrate, heating for 5 min took place with a watch glass cover. The sample was then kept under atmospheric conditions for 24 h, after which the chloride extract was filtered with a filtering assembly. The chloride content was measured using a potentiometric titration apparatus and the water-soluble chloride content in the specimen was quantified.

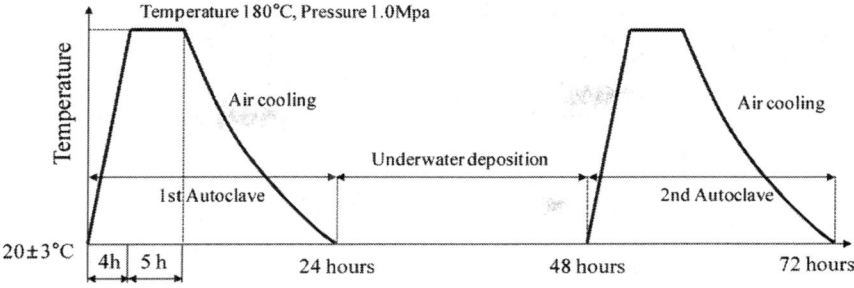

Figure 4: Temperature profile of autoclave-accelerated corrosion.

Figure 5: Procedures for testing the anti-corrosion characteristics. (a) Measurement of the corrosion potential; (b) Accelerated corrosion by autoclave

TEST RESULTS

Electrochemical Anti-Corrosion Characteristics of Corrosion Inhibitors

The test results are listed in Table 7. It is known from previous studies that corrosion is initiated if the molar ratio exceeds the value of 0.6, regardless of the type of corrosion inhibitor. However, the results of this study indicated that the lithium nitrite inhibitor most strongly exhibited corrosion-protection properties at a molar ratio of 0.6 and above, whereas the amino alcohol inhibitors exhibited corrosion protection properties starting at a molar ratio of 0.3. From these results, it was determined that the anti-corrosion properties of an amino alcohol inhibitor were superior to those of a lithium nitrite inhibitor at the same molar ratio.

Table 7: Electrochemical measurement results of reinforcing steel specimen in solution. CSE: copper–copper sulfate electrode

No.	Content of Cl− (kg/m3)	Lithium Nitrite Inhibitor				Amino Alcohol Inhibitor			
		LiNO2				(CH3)2NCH2CH2OH			
		Ecorr	Icorr	CR	Rp	Ecorr	Icorr	CR	Rp
		mV. CSE	μA/cm2	mpy	Ω cm2	mV. CSE	μA/cm2	mpy	Ω cm2
1	0.0	−207	0.092	0.20	282.6	−207	0.092	0.20	282.6
2	0.6	−359	0.98	1.22	26.53	−359	0.98	1.22	38.81
3		−350	0.73	0.98	35.67	−312	0.22	0.28	100.00
4		−333	0.46	0.77	39.16	−309	0.21	0.22	118.18
5		−283	0.19	0.44	63.41	−208	0.11	0.13	166.40
6	1.2	−381	1.19	1.33	21.85	−381	1.19	1.33	21.85
7		−358	0.87	1.02	29.89	−298	0.20	0.35	106.56
8		−311	0.55	0.85	47.27	−291	0.19	0.24	122.64
9		−318	0.20	0.32	81.25	−290	0.18	0.14	144.44
10	2.4	−430	1.26	1.55	20.63	−430	1.26	1.55	20.63
11		−368	1.05	1.45	24.76	−324	0.21	0.27	83.60
12		−340	0.45	0.88	57.78	−278	0.11	0.22	146.09
13		−338	0.22	0.62	86.67	−262	0.10	0.14	170.50
14	4.8	−438	1.50	1.76	17.33	−438	1.5	1.76	17.33
15		−388	0.90	1.54	28.89	−360	0.33	0.19	60.19
16		−344	0.86	0.94	30.23	−311	0.19	0.17	95.94
17		−327	0.23	0.43	49.06	−302	0.18	0.14	123.81

Corrosion Potential for Different Types of Corrosion Inhibitors

Figure 6 shows the relationship between the corrosion potentials at different molar ratios of inhibitor and chloride concentrations for the lithium nitrite and amino alcohol inhibitors. One can see from Figure 6 that in the case of no inhibitor, the corrosion potential increases as the content of Cl^- increases and once the inhibitor is added, the corrosion potential decreases. It is also evident that at the same content of Cl^- and similar molar ratio in both inhibitors, the corrosion potential achieved with DMEA is lower than that of lithium nitrite. This observation is noticed at all concentrations of Cl^-.

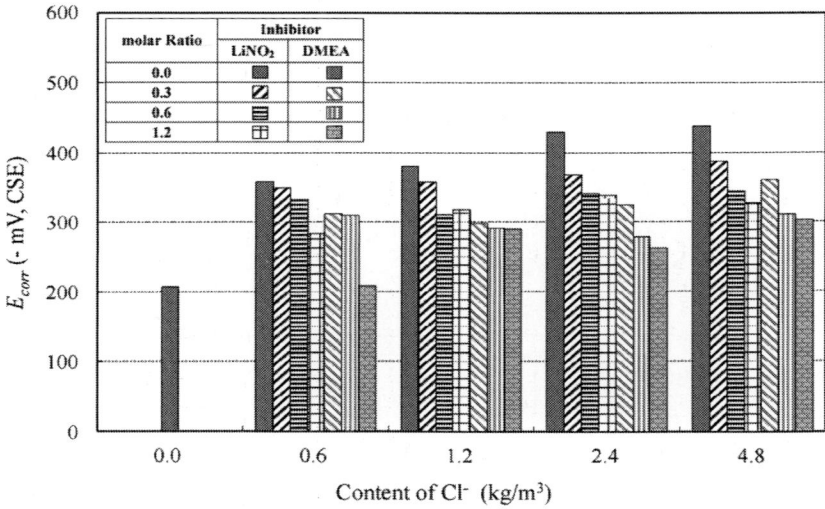

Figure 6: Corrosion potential of the different inhibitors.

Corrosion Current Density for Different Types of Corrosion Inhibitors

Figure 7 shows the relationship between corrosion current density values at different molar ratios of lithium nitrite and amino alcohol inhibitors and concentration of Cl^-. The corrosion current densities were measured at the initiation of reinforcing steel corrosion. Under the presumption that the corrosion rate is considered low when corrosion current density value is in the range 0.2–0.5 $\mu A/cm^2$ [28], both the lithium nitrite and amino alcohol inhibitors showed decreasing corrosion current densities as the molar ratio of the inhibitor increases. Moreover, as is evident from Figure 7, if the corrosion status is evaluated on the basis of the corrosion current density, in the case of the lithium nitrite inhibitor, the corrosion current density was distributed in the range of 0.2–1.0 µA/cm², depending on the chloride ion content and the molar ratio of the inhibitor. Anti-corrosion properties (corrosion rate is passive) are present at chloride ion levels of 0.6 kg/m³ and 1.2 kg/m³and at the molar ratio of inhibitor of 1.2. At chloride ion levels of 2.4 kg/m³ and 4.8 kg/m³, corrosion was low at molar ratios of inhibitor of ≥1.2.

On the other hand, for the amino alcohol inhibitor with a corrosion current density of up to 0.2 µA/cm², anti-corrosion properties (corrosion rate is passive) are present at chloride ion levels of 0.6 kg/m³ and 1.2 kg/m³ and at the molar ratio of inhibitor of 1.2. At chloride ion levels of 2.4 kg/m³ and 4.8 kg/m³, corrosion was low at molar ratios of inhibitor to chloride of ≥0.6, whereupon anti-corrosion properties are clearly evident.

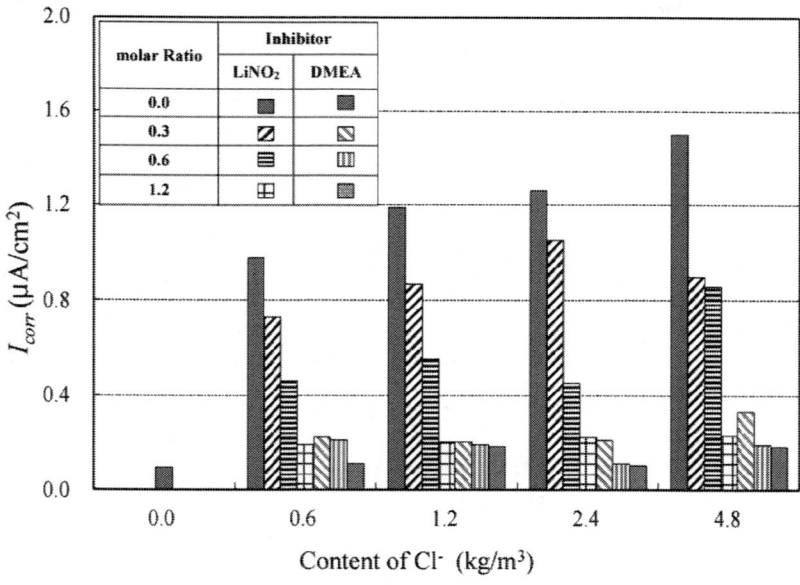

Figure 7: Corrosion current density of the different inhibitors.

Corrosion Rates for the Different Types of Corrosion Inhibitors

Figure 8 shows the relationship between corrosion rate and concentration of Cl^- for lithium nitrite and amino alcohol inhibitors at different molar ratios of inhibitor, in which the anti-corrosion capability was evaluated; it should be below the anti-corrosion standard of 0.5 mpy (dashed horizontal line in Figure 8). The lithium nitrite inhibitor, $LiNO_2$, showed a decrease in corrosion rate with increasing inhibitor molar ratio. In particular, only if the inhibitor molar ratio was ≥ 1.2, the anti-corrosion standard of 0.5 mpy was satisfied. Likewise, the corrosion rate of the amino alcohol inhibitor tended to decrease as the inhibitor molar ratio increased. In particular, the corrosion rate of DMEA quickly dropped below 0.5 mpy in all cases in which the inhibitor molar ratio was ≥ 0.3, regardless of chloride ion levels of 0.6 kg/m^3, 1.2 kg/m^3, 2.4 kg/m^3, and 4.8 kg/m^3. These results indicate that a small amount of amino alcohol corrosion inhibitor can achieve outstanding anti-corrosion capabilities.

Therefore, from the analysis results of both corrosion current density and corrosion rate, the lithium nitrite inhibitor achieved anti-corrosion properties at a molar ratio of approximately 1.2 or greater, whereas the amino alcohol inhibitor achieved anti-corrosion properties at a molar ratio above 0.3. Considering that all results were obtained in an aqueous solution containing chloride ions, even small amounts of amino alcohol inhibitor show better anti-corrosion capabilities than the lithium nitrite inhibitor.

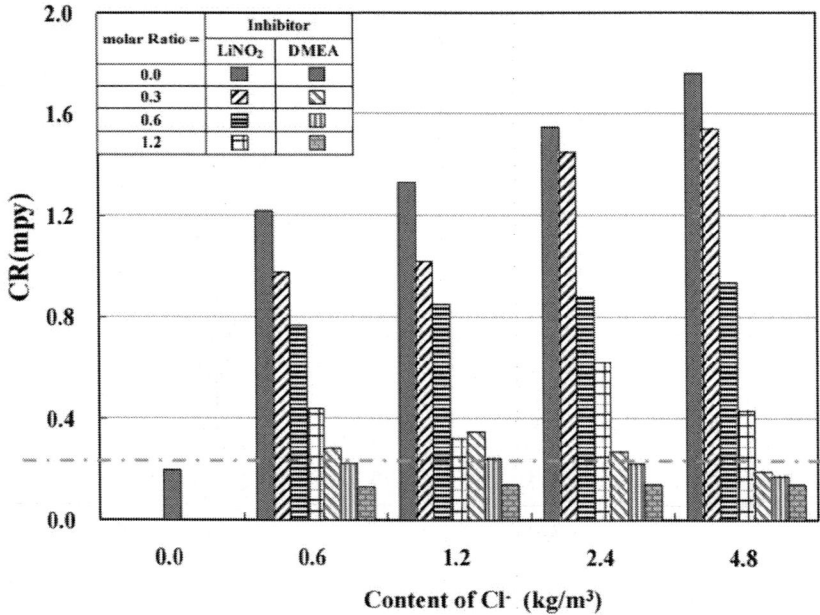

Figure 8: Relationship between corrosion rate and content of Cl⁻.

Polarization Resistances for the Different Types of Corrosion Inhibitors

Figure 9 shows the relationship between the polarization resistance and content of Cl⁻ at different molar ratios of lithium nitrite and amino alcohol inhibitors. The polarization resistance of the lithium nitrite inhibitor, $LiNO_2$, tended to increase with an increase in the molar ratio of the lithium nitrite inhibitor.

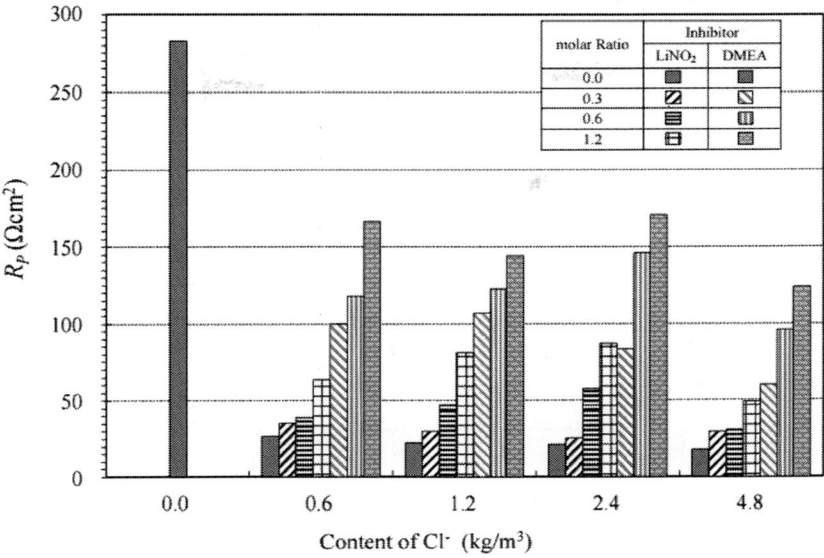

Figure 9: Relationship between polarization resistance and molar ratio.

Thus, it is established that polarization resistance is not an appropriate indicator of anti-corrosion properties. The reason for this is believed to be the nonlinearity of the slopes of the anodic and cathodic polarization curves, which indicates that varying values of the corrosion current density I_{corr} may appear, depending on the detailed test conditions. On the other hand, the polarization resistance of the amino alcohol inhibitors also tended to increase with an increasing molar ratio of inhibitor to chloride.

Testing of Anti-Corrosion Properties of Reinforced Concrete Specimen by the Accelerated Corrosion Method

Table 8 shows the experimental test results of the reinforced concrete specimen that were subjected to accelerated corrosion tests. Figure 10 shows the corrosion in reinforcing steel after being taken from the reinforced concrete specimen. A reinforced concrete specimen was subjected to a chloride ion content of 2.4 kg/m³ and a molar ratio of inhibitor of 1.2. Table 9 displays the corrosion conditions of the reinforcing steel after accelerated corrosion relative to the amount of corrosion inhibitor added. As can be seen from the test results in Table 8, as the molar ratio of the inhibitor increased—in other words, as the amount of added corrosion inhibitor increased—the half-cell potential and the corrosion/area ratio rapidly decreased, clearly indicating the corrosion-inhibiting effects.

Table 8: Experimental test results of the reinforced concrete specimen CSE: copper–copper sulfate electrode

No.	Content of Cl– (kg/m3)	Molar Ratio	Lithium Nitrite Inhibitor			Amino Alcohol Inhibitor		
			LiNO2			(CH3)2NCH2CH2OH		
			Half-Cell Potential	Corrosion Area Ratio	Water-Soluble Cl–	Half-Cell Potential	Corrosion Area Ratio	Water-Soluble Cl–
			mV. CSE	%	%	mV. CSE	%	%
1	0.0	0.0	−211	11.6	0.0037	−211	11.6	0.0037
2	1.2	0.0	−387	68.5	0.0148	−387	68.5	0.0148
3		0.3	−369	34.0	0.0188	−263	13.9	0.0071
4		0.6	−343	30.6	0.0217	−213	13.1	0.0037
5		1.2	−289	18.8	0.0185	−180	6.3	0.0012
6	2.4	0.0	−443	70.6	0.0275	−443	70.6	0.0275
7		0.3	−384	34.3	0.0282	−317	21.1	0.0084
8		0.6	−352	31.5	0.0297	−229	11.5	0.0044
9		1.2	−317	13.8	0.0239	−215	8.0	0.0024

Considering the half-cell potential values and the probability of corrosion as per ASTM C 876-09 [29], the lithium nitrite inhibitor, $LiNO_2$, did not achieve anti-corrosion properties as per the above mentioned standard, as the half-cell potential values were all below −350 mV; this indicates a high corrosion rate and the probability of more than 90% corrosion occurring. However, the data also show that adding lithium nitrite helped in reducing the half-cell potential values measured, compared to cases where no inhibitor was added.

On the other hand, the amino alcohol inhibitor, DMEA, showed obvious anti-corrosion properties starting from a molar ratio of inhibitor of 1.2. Also, in terms of the water-soluble chloride content, which has a direct impact on reinforcing steel corrosion, the lithium nitrite inhibitor showed a slight decrease, whereas amino alcohol inhibitors showed a drastic decrease of the chloride content with an increasing amount of the added corrosion inhibitor. These results indicate that the amino alcohol inhibitor has a superior anti-corrosion effect in a saline environment because of its chloride ion binding property.

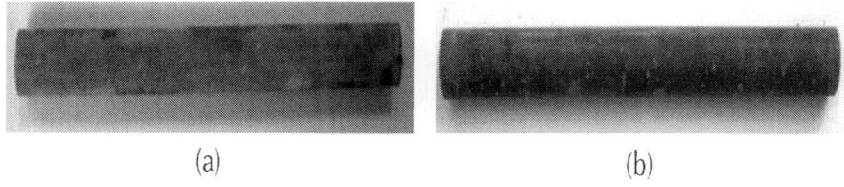

(a)   (b)

Figure 10: Appearance of reinforcing steel corrosion at the chloride ion concentration of 2.4 kg/m³ and inhibitor molar ratio of 1.2. (a) Lithium nitrite inhibitor (LiNO₂); (b) Amino-alcohol inhibitor (DMEA).

Table 9: Corrosion states of the reinforcing steel after accelerated corrosion

Content of Cl−	Inhibitor Molar Ratio	Corrosion Area of Rebar	
		LiNO2	DMEA
0.0 kg/m3	0.0		
1.2 kg/m3	0.0		
	0.3		
	0.6		
	1.2		
2.4 kg/m3	0.0		
	0.3		
	0.6		
	1.2		

Half-Cell Potentials of the Reinforced Concrete Specimens for Different Corrosion-Inhibitor Types

Figure 11 shows the relationship between half-cell potential values and content of Cl⁻ at different molar ratios of lithium nitrite and amino alcohol inhibitors. For all investigated inhibitors, it was confirmed that as the molar ratio of the inhibitor increased, the half-cell potential negative values decreased. As indicated in Figure 11, considering only half-cell potentials of ≤−200 mV, the standard level for achieving anti-corrosion properties (ASTM C 876-09) [29], the lithium nitrite inhibitor, $LiNO_2$, did not achieve anti-corrosion capabilities at any of the molar ratio used. On the other hand, the amino alcohol inhibitor immediately demonstrated anti-corrosion properties at the two chloride ion concentrations of 1.2 kg/m³ and 2.4 kg/m³ for a molar ratio of inhibitor of 1.2.

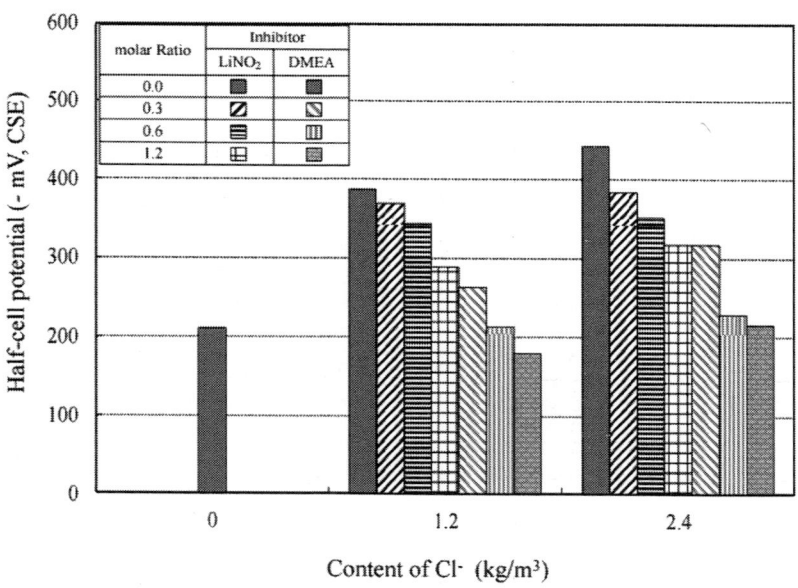

Figure 11: Corrosion potential of the rebar at different molar ratios of inhibitor.

Corrosion Areas of Reinforcing Steel in Reinforced Concrete Specimens for the Different Corrosion-Inhibitor Types

Figure 12 displays the corrosion-area ratios related to the molar ratio of lithium nitrite and amino alcohol inhibitors. As evidenced by Figure 12, the amino alcohol inhibitor has superior anti-corrosion properties because the corrosion area is smaller when the amino alcohol rather than the lithium nitrite inhibitor is added.

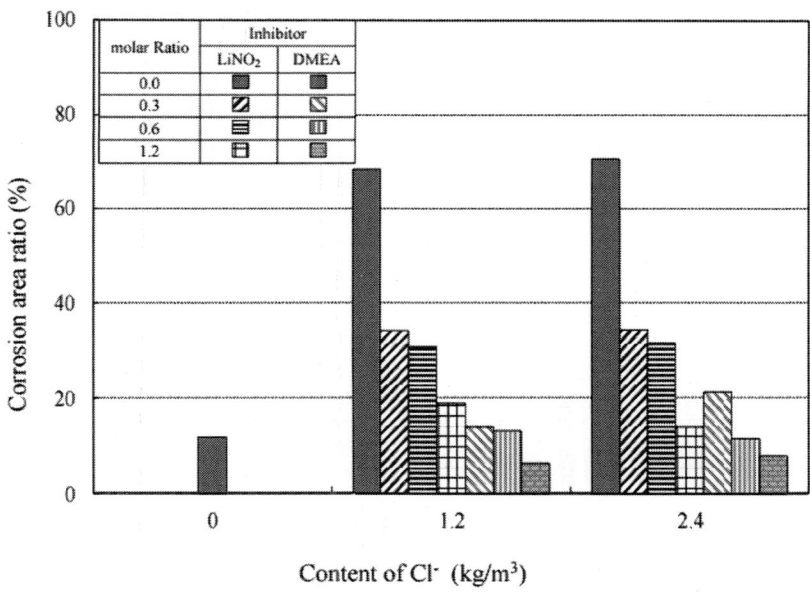

Figure 12: Corrosion/area ratio at various molar ratios of inhibitor.

In particular, at a chloride ion concentration of 1.2 kg/m³ and a molar ratio of inhibitor of 0.3, a corrosion/area ratio of 34.0% was observed in cases where the lithium nitrite inhibitor was added, in comparison to a corrosion/area ratio of 13.9% for the amino alcohol inhibitor under the same conditions, which represents a 2.5-fold decrease in the corrosion area.

Additionally, when using the amino alcohol inhibitor at chloride ion concentrations of 1.2 kg/m³ and 2.4 kg/m³, as the molar ratio of the inhibitor increased, corrosion drastically decreased, which indicates that its corrosion-prevention capabilities were much better than those of the lithium nitrite inhibitor

Chlorides in Hardened Concrete for Different Corrosion-Inhibitor Types

Figure 13 shows the water-soluble chloride content in hardened concrete at different molar ratios of lithium nitrite and amino alcohol inhibitors. In terms of the soluble chloride levels, which have a direct impact on reinforcing steel corrosion, the water-soluble chloride content only slightly varied as the added amount of the lithium nitrite corrosion inhibitor increased. In contrast, in cases where the amount of the amino alcohol inhibitor increased, the water-soluble chloride content showed a rapid decrease. Particularly, this rapid decrease in chloride content was accomplished already at a molar ratio of inhibitor to chloride of 0.3, where the anti-corrosion effects of the amino alcohol inhibitor in saline environments are superior to those of the lithium nitrite inhibitor due to the binding effect with respect to chloride ions.

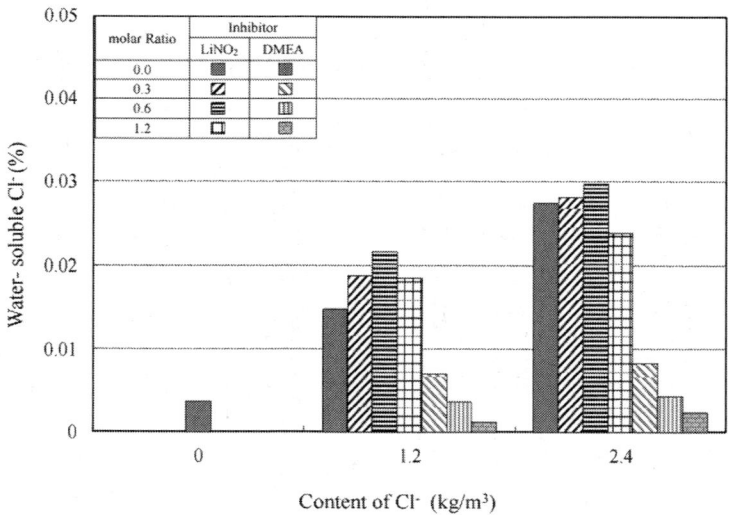

Figure 13: Relationship between water-soluble chloride content and content of Cl⁻ added at different molar ratios of corrosion inhibitor.

CONCLUSIONS

Based on the present experimental results, the following conclusions are drawn:

- Based on the half-cell potential results, the amino alcohol inhibitor was evaluated to have superior anti-corrosion capabilities compared with the lithium nitrite inhibitor.

- The results of this study indicated that the lithium nitrite inhibitor most strongly exhibited corrosion-protection properties at a molar ratio of 0.6 and above, whereas the amino alcohol inhibitors exhibited corrosion-protection properties starting at a molar ratio of 0.3.

- The lithium nitrite inhibitor secured anti-corrosion capacity at chloride ion levels of 2.4 kg/m^3 and 4.8 kg/m^3; corrosion was low at molar ratio of inhibitor of ≥1.2. Amino alcohol inhibitor DMEA showed also anti-corrosion capacity at chloride ion levels of 2.4 kg/m^3 and 4.8 kg/m^3; corrosion was low at molar ratios of inhibitor to chloride of ≥0.6.

- Water-soluble chloride content only slightly varied as the amount of the lithium nitrite corrosion inhibitor increased. In contrast, in cases where the amount of the amino alcohol inhibitor increased, the water-soluble chloride content showed a rapid decrease. Particularly, this rapid decrease in chloride content was accomplished already at a molar ratio of inhibitor to chloride of 0.3.

- Through the effect of binding chloride ions, the amino alcohol inhibitor showed anti-corrosion effects in saline environments that were clearly superior to those of the lithium nitrite inhibitor.

ACKNOWLEDGMENTS

This work was supported by the research fund of Hanyang University (HY-2013-P).

AUTHOR CONTRIBUTIONS

Hwa-Sung Ryu and Won-Jun Park conducted the experiments and wrote the initial draft of the manuscript. Han-Seung Lee designed the project and Hwa-Sung Ryu analyzed the data and wrote the final manuscript. Mohamed A. Ismail reviewed and contributed to the final revised manuscript. All authors contributed to the analysis of the data and read the final paper.

REFERENCES

1. Song, H.W.; Lee, C.H.; Ann, K.Y. Factors influencing chloride transport in concrete structures exposed to marine environment. *Cem. Concr. Compos.* 2008, *30*, 113–121.

2. Song, H.W.; Ann, K.Y. Chloride threshold level for corrosion of steel in concrete. *Corros. Sci.* 2007, *49*, 4113–4133.

3. Glass, G.K.; Buenfeld, N.R. The presentation of the chloride threshold level for corrosion of steel in concrete. *Corros. Sci.* 1997, *39*, 1003–1013.

4. Song, H.W.; Lee, C.H.; Lee, K.C.; Ann, K.Y. A Study on chloride threshold level of blended cement mortar using polarization resistance method. *J. Korea Concr. Inst.* 2009, *21*, 245–253.

5. Cho, H.K.; Yoo, J.H.; Lee, H.S. Experimental study on the penetration depth and concentration of corrosion inhibitor using press-in method into the inside of concrete. *J. Korea Inst. Struct. Maint. Insp.* 2009, *13*, 160–168.

6. Kwon, S.J.; Song, H.W.; Byun, K.J. A Study on analysis technique for chloride penetration in cracked concrete under combined deterioration. *J. Korea Concr. Inst.* 2007, *19*, 359–366.

7. Moon, B.C.; Yoo, J.H.; Lee, H.S.; Lee, M.S. An experimental study on the development of mortar setting time control of nitrate corrosion inhibitor and the evaluation of corrosion protection effect for electrochemical. *J. Arch. Inst. Korea* 2009, *25*, 113–120.

8. Bolzoni, F.; Lazzari, L.; Ormellese, M.; Goidanich, S. Prevention of corrosion in concrete: The use of admixed inhibitors. In Proceedings of the CORROSION NACExpo 2006, San Diego, CA, USA, 12–16 March 2006.

9.   Chambers, B.D.; Taylor, S.R.; Lane, D.S. *An Evaluation of New Inhibitors for Rebar Corrosion in Concrete*; Final Report. Virginia Transportation Research Council: Charlottesville, VA, USA, 2003.

10. Mennucci, M.M.; Costa, I. Evaluation of benzotriazole as corrosion inhibitor for carbon steel as reinforcement of concrete. In Proceedings of the 210th Meeting of the Electrochemical Society, Cancun, Mexico, 29 October–3 November 2006.

11. *Accelerated Corrosion Test Method of Reinforced Concrete— Part 1: Autoclave Method*; Korean Standard KS F 2599-1. Korean Agency for Technology and Standards (KATS): Seoul, Korea, 2013.

12. Tae, S.H.; Kim, R.H. A study on improvement of durability of reinforced concrete structures mixed with natural inorganic minerals. *Constr. Build. Mater.* 2011, *25*, 4263–4270.

13. Tae, S.H. Corrosion inhibition of steel in concrete with natural inorganic minerals in corrosive environments due to chloride attack. *Constr. Build. Mater.* 2012, *35*, 270–280.

14. Chen, D.; Yen, M.; Lin, P.; Groff, S.; Lampo, R.; McInerney, M.; Ryan, J. A corrosion sensor for monitoring the early-stage environmental corrosion of A36 carbon steel. *Materials* 2014, *7*, 5746–5760. [Google Scholar] [CrossRef]

15. Ismail, M.A.; Soleymani, H.; Ohtsu, M. Early detection of corrosion activity in reinforced concrete slab by AE technique. In Proceedings of the 6th Asia-Pacific Structural Engineering and Construction Conference (APSEC 2006), Kuala Lumpur, Malaysia, 5–6 September 2006.

16. Ismail, M.; Ohtsu, M. Corrosion rate of ordinary and high-performance concrete subjected to chloride attack by AC impedance spectroscopy. *Constr. Build. Mater.* 2006, *20*, 458–469.

17. So, H.S. Environmental influences and assessment of corrosion rate of reinforcing bars using the liner polarization resistance technique. *J. Arch. Inst. Korea* 2006, *22*, 107–114.

18. Saraswathy, V.; Song, H.W. Performance of galvanized and stainless steel rebars in concrete under macrocell corrosion conditions. *Mater. Corros.* 2005, *56*, 685–691.

19. Lee, H.S.; An, J.M.; Shin, S.W. Study on the evaluation of corrosion protection effect with $LiNO_2$, inhibitor. *J. Korea inst. Struct. Maint. Insp.* 2002, *6*, 339–342.

20. Gaidis, J.M. Chemistry of corrosion inhibitors. *Cem. Concr. Compos.* 2004, *26*, 181–189.

21. Ormellese, M.; Berra, M.; Bolzoni, F.; Pastore, T. Corrosion inhibitors for chlorides induced corrosion in reinforced concrete structures. *Cem. Concr. Res.* 2006, *36*, 536–547.

22. Duprat, M.; Bui, N.; Dabosi, F. Corrosion inhibition of a carbon steel by 2-ethylamino-ethanol in aerated 3% NaCl solutions—Effect of pH. *Corrosion* 1979, *35*, 392–397.

23. Duprat, M.; Dabosi, F. Corrosion inhibition of a carbon steel in 3% NaCl solutions by aliphatic amino-alcohol and diamine type compounds. *Corrosion* 1981, *37*, 89–92.

24. Maeder, U. A new class of corrosion inhibitors for reinforced concrete. In Proceedings of the 9th Asian-Pacific Corrosion Control Conference, "Corrosion Prevention for Industrial Safety and Environmental Control", Kaohsiung, Taiwan, 5–10 November 1995; pp. 825–830.

25. Dean, S.W. Electrochemical methods of corrosion testing. In *Electrochemical Techniques for Corrosion*; Baboian, R., Ed.; NACE: Houston, TX, USA, 1977; pp. 52–60.

26. Stern, M.; Geary, A.L. Electrochemical polarization, No. 1 theoretical analysis of the shape of polarization curves. *J. Electrochem. Soc.* 1957, *104*, 56–57.

27. *Standard Test Method for Water-Soluble Chloride in Mortar and Concrete*; ASTM C 1218/C1218M-99 (Reapproved 2008). American Society for Testing and Materials: West Conshohocken, PA, USA, 2008.

28. Building Research Establishment (BRE). Corrosion of Steel in Concrete. In *Investigation and Assessment*; BRE: London, UK, 2001.

29. *Standard Test Method for Half-Cell Potentials of Uncoated Reinforcing Steel in Concrete*; ASTM C876-09. American Society for Testing and Materials: West Conshohocken, PA, USA, 2008.

Chapter 7

A Corrosion Sensor for Monitoring the Early-stage Environmental Corrosion of A36 Carbon Steel

Dong Chen[1], Max Yen[1], Paul Lin[1], Steve Groff[1], Richard Lampo[2], Michael McInerney[2], and Jeffrey Ryan[2]

[1]College of Engineering, Technology, and Computer Science, Indiana University-Purdue University, 2101 E Coliseum Blvd, Fort Wayne, IN 46805, USA;

[2]U.S. Army Engineer Research & Development Center, Construction Engineering Research Laboratory, P.O. Box 9005, Champaign, IL 61826, USA;

ABSTRACT

An innovative prototype sensor containing A36 carbon steel as a capacitor was explored to monitor early-stage corrosion. The sensor detected the changes of the surface- rather than the bulk- property and morphology of A36 during corrosion. Thus it was more sensitive than the conventional electrical resistance corrosion sensors. After being soaked in an aerated 0.2 M NaCl solution, the sensor's normalized electrical resistance (R/R_0) decreased continuously from 1.0 to 0.74 with the extent of corrosion. Meanwhile, the sensor's normalized capacitance (C/C_0) increased continuously from 1.0 to 1.46. X-ray diffraction result indicates that the iron rust on A36 had crystals of lepidocrocite and magnetite.

INTRODUCTION

Corrosion is a destructive attack on a metal such as carbon steel, aluminum, zinc and copper by chemical or electrochemical reactions with its environment [1]. It is a spontaneous process. If corrosion is not monitored and correctly fixed, it could threaten public welfares and people's lives [2]. Among various causes of corrosion, environmental factors are the most common ones because of ubiquitousness. Practically all environments are corrosive to some degree [3]. Some examples are air and moisture; fresh, distilled, salt, and mine waters; steam and other gases such as chlorine, ammonia, hydrogen sulfide, and fuel gases; mineral and organic acids [3,4]. Among these environmental factors, chloride is an important one. It is well known that chloride ions can cause passive layer breakdown and corrosion of metals [5,6]. Structures can be exposed to chloride ions through various means including deicing salts, fresh water, and a marine environment [7].

Manual inspection of corrosion is costly, low efficient, subjective and sometimes dangerous. It typically requires a large amount of time for professionals to travel and inspect each site. Especially when there are difficult-to-access or completely inaccessible areas, manual inspections are almost impossible. As a result, it is highly desirable to use corrosion sensors for automatic data collection, processing, and evaluation. Compared to manual inspections, automatic monitoring by corrosion sensors has significant advantages, such as promptness, comprehensiveness and efficiency. In addition, electrical signals from corrosion sensors are much easier to transmit, analyze and store than manual methods.

Conventional corrosion sensors are typically based on the mechanism of an increase in electrical resistance of iron with the degree of corrosion [8,9]. However, a lengthy response time is required to register a significant change in corrosion rate [10], because the percentage of the thickness change of the sensor has to be noteworthy.

Detection of the early-stage environmental corrosion is of critical importance to maintain the integrity and the safety of structures and systems, because the corrosion can be a self-accelerating process when no corrosion inhibitors are present [11, 12]. As a result, a prototype corrosion sensor has been explored in this study, in order to find the sensitive and systematic change in electrical properties of a metal surface (e.g., A36 carbon steel as investigated in this study) during the early-stage corrosion as it is exposed to a corrosive environment. A36 carbon steel is commonly used in steel bridges and other structures [13, 14]. The definition of the early-stage corrosion is the mass change of the sensor is within 0.2% (or iron loss per exposed surface area is within 187.7 g/m^2) as explored in this study. The prototype sensor was essentially a capacitor composed of two parts: (i) the same metal with the same passivation/coating as the metallic structure or the system to be monitored; (ii) a corrosion-resistant conductor. The two parts were separated by air, the same corrosion environment of the structure or the system in service. As a result, the corrosion of the sensor represents the extent of corrosion of the structure/system being monitored. During the course of corrosion and degradation, there is a rapid change of the surface morphology and the property of the metal, rather than a slow change of the bulk electrical resistance measured by the conventional corrosion sensors. Therefore, the degree of corrosion can be sensitively reflected by the systematic change of the capacitance and the resistance readings from the prototype sensor. To our best knowledge, this type of corrosion sensor based on surface electrical resistance and capacitance measurements has not been reported yet.

In practice, the sensor can be connected to a wired or wireless network for automatic data acquisition, processing and storage. Multiple sensors can be deployed at varied locations of a structure to provide a comprehensive monitoring network without the need for a site visitation. Thus it is more efficient and cost-saving. This can make it possible to remotely monitor the extent of corrosion of a structure or a system that the sensors are attached to. Figure S1 and S2 in the Supporting Information show an example of the installed corrosion sensors from this study on a steel bridge, the data acquisition and the monitoring system. The sensors can be combined with conventional bulk electrical resistance sensors, which are not sensitive to early-stage corrosion. Thus a monitoring system is formed to examine corrosion of varied stages. The main objective of this study was to develop a sensor system that could sensitively determine the degree of the early-stage corrosion of steel and steel structures during service; thereby, giving a chance to estimate the integrity of the infrastructures and to apply corrosion-control measures timely, so that a catastrophic failure could be prevented.

RESULTS AND DISCUSSION

Iron Loss during Corrosion

A new cylindrical corrosion sensor consisted of a rust-free A36 carbon steel rod in the centre and the 316 stainless steel ring (see the section of Experimental Procedures). Two electrical wires were connected to them respectively. During the corrosion test, rust was visually observed on the A36 steel as early as 2 h of exposure to an aerated 0.2 M NaCl solution. In the meantime, the NaCl solution turned yellowish in colour with suspended small rust particles. After accumulated 225.5 h in an aerated 0.2 M NaCl solution, rust was very apparent and had covered a large surface area of the A36 steel rod. However, as expected, no visible corrosion was observed on the 316 stainless steel ring or the stainless steel reference sensor.

During the corrosion process, yellowish iron rust continuously released from the sensor to the NaCl solution. The amount of iron in the solution was quantified by Atomic Absorption Spectroscopy (AAS) after dissolution of the rust with 10% (v/v) nitric acid. The results indicate that 0.16–0.81 mg/day of iron was released from the sensor to the solution. As shown in Figure 1, at the end of the test of 225.5 h, 3.24 mg of the accumulated iron was in the solution. The corresponding corrosion rate varied between 0.60 and 3.02 g/(m^2·day). It should be noted that there was rust on the A36 steel surface in addition to the amount found in the NaCl solution. The rust was not cleaned from the sensor intentionally to maintain its natural condition during corrosion process. As a result, the overall iron loss and corrosion rate should be greater than that found in the solution as shown in Figure 1. The varied corrosion rate in this study is likely due to different amounts of rust spalling from the sensor from time to time, which can affect the quantity of iron in the solution significantly. However, for the control test of the 316 stainless steel ring alone or the stainless steel reference sensor, the dissolved iron concentration was less than 0.02 mg at the end of 225.5 h of corrosion, indicating the corrosion was insignificant.

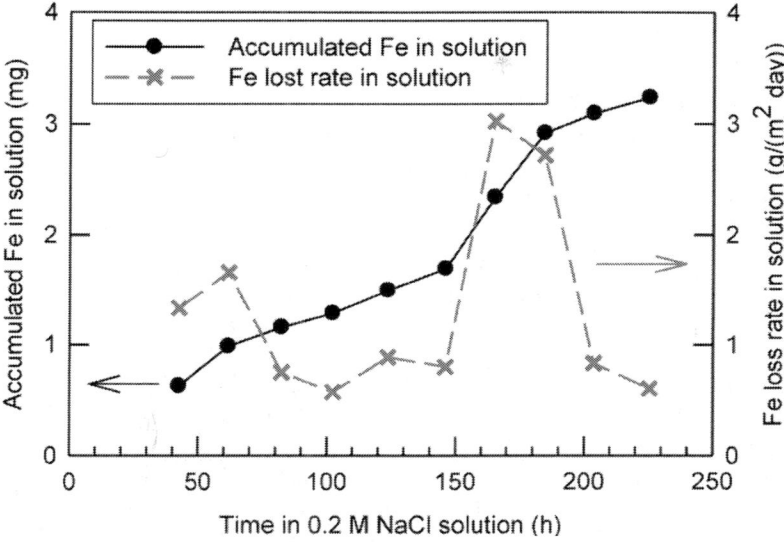

Figure 1: The iron loss rate and the accumulated iron loss in the solution during the test of the prototype corrosion sensor containing A36 steel in an aerated 0.2 M NaCl solution.

Despite apparent corrosion of the A36 steel rod through AAS measurements and visual observations, the mass of the sensor was maintained almost constant at 25.20 ± 0.05 g (*i.e.*, variation was within $\pm 0.2\%$) throughout the test. This result is consistent with Figure 1, in which the iron loss in the solution was within mg range. Based on the reactions (1) to (9), the loss of iron (Fe) from the sensor can be compensated by gains of oxygen and hydrogen atoms in the rust. As a result, mass, as a bulk parameter is not sensitive to evaluate corrosion. On the other hand, the minor change of the mass suggests the early-stage corrosion. The air gap distance between the A36 steel rod and the 316 stainless steel ring of the sensor did not increase significantly, which is an important evidence for the explanation of the electrical resistance and the capacitance changes of the sensor during corrosion in later discussions.

Compositions of the Rust

The corrosion of carbon steel can occur as an electrochemical reaction, with one anodic reaction and one cathodic reaction [3,15]. The anodic reaction usually occurs as:

$$Fe \rightarrow Fe^{2+} + 2e^- \qquad (1)$$

The cathodic reaction, however, can be different depending on what is in the environment. This cathodic reaction is the main factor that influences the rate of corrosion [3,15]. In an aerated solution it is most likely:

$$O_2 + 2H_2O + 4e^- \rightarrow 4OH^- \qquad (2)$$

Combining the cathodic and anodic reactions gives:

$$2Fe + O_2 + 2H_2O \rightarrow 2Fe(OH)_2 \downarrow \qquad (3)$$

As a result, ferrous hydroxide precipitates from the solution. However, dissolved oxygen can oxidize ferrous hydroxide to ferric hydroxide:

$$4Fe(OH)_2 + O_2 + 2H_2O \rightarrow 4Fe(OH)_3 \downarrow \qquad (4)$$

Yellowish/brownish rust was observed at the A36 carbon steel surface and in the solution. In addition, black rust was also seen on the A36 steel underneath the yellowish/brownish rust, which is likely magnetite (Fe_3O_4). The formation of magnetite is due to iron not having enough oxygen present for the reaction [16]. The following additional reactions may occur involving oxidation of iron and producing rusts at the A36 carbon steel surface [15,17,18].

$$Fe^{2+} \rightarrow Fe^{3+} + e^- \qquad (5)$$

$$2Fe(OH)_2 \rightarrow Fe_2O_3 + H_2O + 2H^+ + 2e^- \qquad (6)$$

$$4Fe(OH)_2 + O_2 \rightarrow 4FeOOH + 2H_2O \qquad (7)$$

$$Fe^{2+} + 8FeOOH + 2e^- \rightarrow 3Fe_3O_4 + 4H_2O$$
(8)

$$2FeO + H_2O \rightarrow Fe_2O_3 + 2H^+ + 2e^-$$
(9)

To examine the crystalline compositions of the iron rust on the A36 steel surface, x-ray diffraction (XRD) analysis was performed with CuKα radiation. Figure 2a shows the XRD spectrum of uncorroded A36 steel. Figure 2b shows the XRD spectrum of the corroded A36 steel soaked in an aerated 0.2 M NaCl solution for 225.5 h. Three types of crystalline substances were found on the corroded steel surface according to their characteristic diffraction patterns [19,20], which were 1: iron, 2: lepidocrocite, and 3: magnetite.

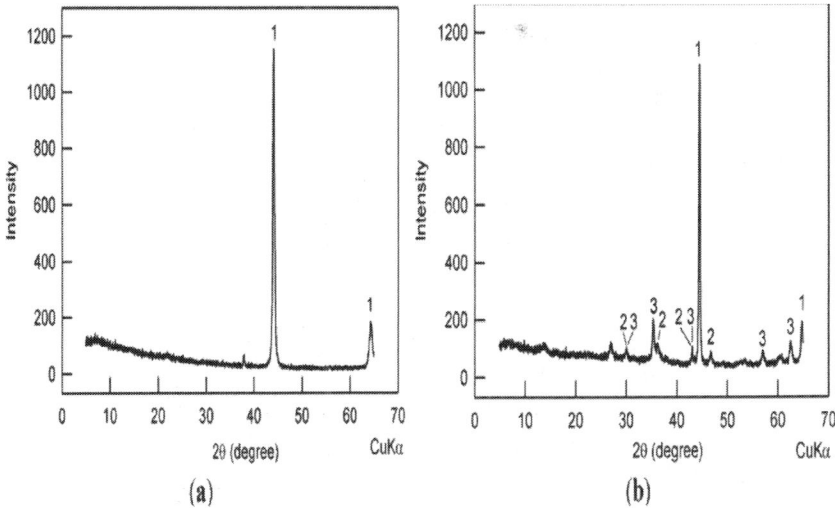

Figure 2: XRD patterns of the (a) uncorroded and (b) corroded A36 steel samples. The identified crystals are 1: iron, 2: lepidocrocite, 3: magnetite.

Consistently, it has been reported that the rust formed on steel surface is a mixture of lepidocrocite (γ-FeOOH), magnetite (Fe_3O_4), hematite (α-Fe$_2$O$_3$), goethite (α-FeOOH), and amorphous iron oxide [21,22,23,24], although only the first two were found in this study. To make the study more general, the resistivity and the dielectric constant of the common rust materials, along with iron and air are listed in Table 1.

Table 1: Electrical resistivity and the dielectric constant of materials related to iron rust at ambient temperature

Materials	α-Fe$_2$O$_3$	γ-FeOOH	Fe$_3$O$_4$	α-FeOOH	Amorphous Fe$_2$O$_3$	Iron	Air
Electrical resistivity ρ ($\Omega \cdot$m)	$(1.58-5.62)$ $\times 10^4$ [25]	$(0.20-0.80)$ $\times 10^5$ [26]	1.58 $\times 10^{-4}$ -0.1 [27]	$(1.30-2.33)$ $\times 10^5$ [26]	2.12×10^3 [28]	1.0 $\times 10^{-7}$	4 $\times 10^{13}$
Dielectric constant ε	12	2.6 [29]	20	11 [30]	4.5	–	1

Electrical Resistance of the Sensor

Figure 3 shows the electrical resistance of the sensor in parallel (*i.e.*, the built-in measurement mode of a resistance, capacitance and inductance (RCL) meter) with the extension of corrosion time in the NaCl solution. In Figure 3, during the course of corrosion, the resistance gradually decreased following an apparent trend. At the end of the corrosion test of 225.5 h, the normalized resistance (R/R_0) decreased from 1.0 to 0.74, a decline of 26%.

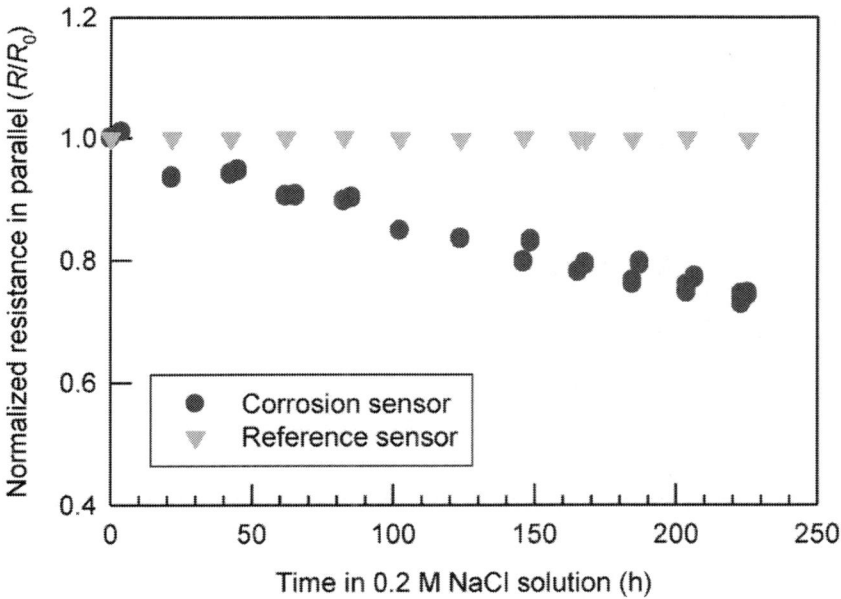

Figure 3: Normalized electrical resistance (R/R_0) of the prototype sensors *vs.* the accumulated time in an aerated 0.2 M NaCl solution.

The electrical resistance for a coaxial cylinder can be expressed by the following equation.

$$R = \frac{\rho}{2\pi h} \ln\left(\frac{b}{a}\right) \tag{10}$$

where R = electrical resistance of a material (Ω); ρ = electrical resistivity of the material ($\Omega \cdot m$);

h = height of the cylinder sensor (m); b= inner radius of the 316 stainless steel ring (m); a = radius of the A36 steel rod (m).

After corrosion, the multi-material resistor (iron rust on A36 steel surface and the air gap) can be treated as resistance in-series to calculate the overall resistance of the sensor,

$$R = \frac{1}{2\pi h}[\rho_1 ln \frac{a + \delta}{a - \delta(1 - \varphi)} + \rho_2 ln \frac{b}{a + \delta}]$$

(11)

where ρ_1 = equivalent electrical resistivity of the porous iron rust on the A36 steel rod surface ($\Omega \cdot m$); ρ_2 = electrical resistivity of the air gap between the iron rust on the A36 steel rod surface and the 316 stainless steel ring ($\Omega \cdot m$); δ = average thickness of the iron rust on the A36 steel rod surface (m); φ = average porosity of the iron rust on the A36 steel rod surface.

The resistance of a new sensor (before corrosion) is mainly due to the air gap between the central cylindrical A36 steel rod and the 316 stainless steel ring (see the section of Experimental Procedures). Air has an electrical resistivity of 4×10^{13} $\Omega \cdot m$ [31]. This is different from the conventional corrosion sensors, which measure the bulk electrical resistance of the metal. When corrosion of the steel rod happened, rust formed a porous and loose structure [22] extended from the A36 steel surface to the surrounding air. As a result, the rust took a partial space that was previously occupied by air. In other words, after corrosion the gap between the A36 steel rod and the 316 stainless steel ring was partially filled with porous rust (small portion) and air (big portion). Equation (11) shows the overall resistance of the sensor as a multi-material resistor (iron rust and air gap). As mentioned earlier, the iron rust is a mixture of lepidocrocite (γ-FeOOH) and magnetite (Fe_3O_4) as shown in Figure 2b, along with reported hematite (α-Fe_2O_3), goethite (α-FeOOH) and amorphous iron oxide [21,22,23,24]. Their electrical resistivity is shown in Table 1. As can be seen, the electrical resistivity of the rust components is at least eight orders of magnitude lower than air. According to Equation (11), a lower electrical resistivity has smaller resistance. Consequently, the electrical resistance decreases with the time or the extent of corrosion. Different from conventional corrosion sensors, which detect an increase in the bulk electrical resistance of iron with corrosion [8, 9], the cylindrical sensor explored in this study examined the A36 surface property and morphology changes. Thus it is more sensitive to monitor the early-stage corrosion. In addition, the sensitivity of the sensor can be fine-tuned by optimizing the values of a and b. The closer the values of a and b, the greater sensitivity of the corrosion sensor is expected to have.

Although spalling of rust from the sensor could enlarge the air gap between the A36 steel rod and the 316 stainless steel ring, the insignificant mass-loss result (i.e., mass change $\leq 0.2\%$ or 187.7 g/m^2) indicates this was not the case during the testing period of early-stage corrosion. However, if significant spalling of rust happens (i.e., $a - \delta(1 - \varphi)$ becomes much small) and thus the air gap between the A36 steel rod and the 316 stainless steel ring increases, the trend of electrical resistivity can be reversed (i.e., electrical resistance increases with time instead), suggesting much severe corrosion, which can be regarded as middle- or late-stage corrosion. This turning point can be used to rank the risk level of corrosion. In contrast, the electrical resistance of the stainless steel reference sensor was stable, as shown in Figure 3. In practice, the reference sensors can be used to normalize the baseline signal of electrical resistances including the effects of air moisture and temperature, in order to distinguish the electrical resistance changes caused by rust formation on the steel surface.

Capacitance of the Sensor

In addition to the electrical resistances, the capacitance of the sensor during corrosion was also examined. Again, the capacitance was from the built-in parallel measurement mode of the RCL meter. Figure 4 shows the change of the capacitance vs. accumulated time of corrosion in an aerated 0.2 M NaCl solution. A positive trend of the capacitance with the extent of corrosion was observed. More specifically, the normalized capacitance (C/C_0) increased from 1.0 at the beginning to 1.46 after 225.5 h. In other words, the capacitance had an increase of 46%.

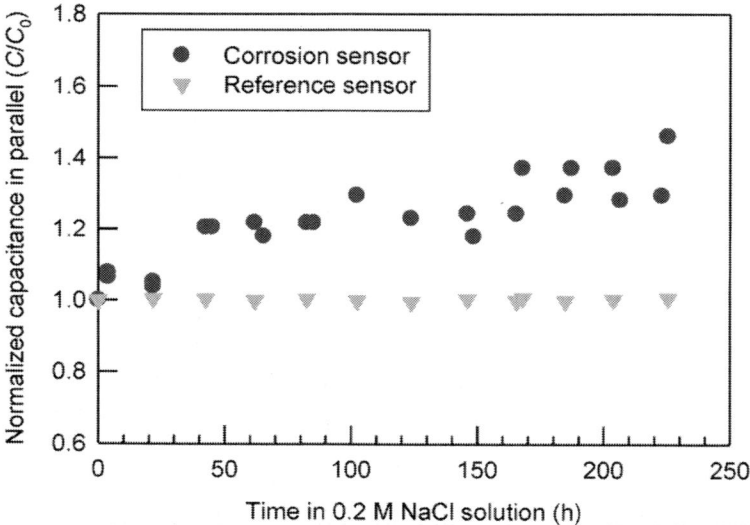

Figure 4: Normalized capacitance (C/C_0) of the prototype sensors vs. the accumulated time in an aerated 0.2 M NaCl solution.

The capacitance of an infinite cylindrical sensor, neglecting the fringing effect, can be calculated from the following equation [32] (see the Supporting Information of a reference equation of the capacitance with fringing effect considered),

$$C = \frac{2\pi\varepsilon_0 h * \varepsilon}{\ln\left(\frac{b}{a}\right)}$$

(12)

where C = capacitance (pF); ε_0 = free space permittivity, 8.85 pF/m; h = height of the cylinder sensor (m); = dielectric constant of the material(s) between the A36 steel rod and the 316 stainless steel ring.

Before corrosion, air was between the A36 steel rod and the 316 stainless steel ring. Air has a dielectric constant of 1 [31]. After corrosion, rust was formed at the A36 steel surface. It means the space between the A36 steel rod and the 316 stainless steel ring was filled with both rust and air, although rust has a much smaller volume than air in this case. The multi-material capacitor (iron rust on A36 steel surface and the air gap) can be treated as capacitors-in-series to calculate the overall capacitance of the sensor,

$$C = 2\pi\varepsilon_0 h \frac{\varepsilon_1 \varepsilon_2}{\varepsilon_1 \ln\left(\frac{b}{a+\delta}\right) + \varepsilon_2 \ln\left(\frac{a+\delta}{a - \delta(1-\varphi)}\right)}$$

(13)

where ε_1 = equivalent dielectric constant of the porous iron rust on the A36 steel rod surface; ε_2 = dielectric constant of the air gap between the iron rust on the A36 steel rod surface and the 316 stainless steel ring.

As corrosion takes place, iron rust gradually grows on the steel surface (i.e., increase in δ from zero initially). As shown in Table 1, the dielectric constant ε_1 of the substances composing the iron rust ranges from 2.6 to 20, much greater than air. Equation (13) shows the overall capacitance C is proportional to ε_1. An increase in ε_1 and/or δ raises the capacitance of the corrosion sensor, although further sensitivity analysis of Equation (13) indicates that C is much more sensitive to ε_1 than δ. However, spalling of the rust from the A36 steel and thus a decrease in $a - \delta(1 - \varphi)$ was insignificant during the early-stage corrosion, because of the measured almost constant mass of the sensor during the test. Consequently, a higher capacitance reading reflects more rust formation at the A36 steel surface of the sensor during the early-stage corrosion. However, if spalling of the rust is significant enough (i.e., $a - \delta(1 - \varphi)$ becomes much small), a decrease in capacitance would indicate severe corrosion, which can be regarded as middle- or late-stage corrosion. Same as the resistance, this turning point of capacitance trend (i.e., decrease in capacitance with time instead) can be used to rank the risk level of corrosion. Again, the reference sensor had little change in capacitance with corrosion time as shown in Figure 4. The reference sensor can be used to normalize the environmental factors (e.g., air moisture and temperature) caused capacitance changes other than corrosion.

The configuration of a short cylindrical sensor (height is comparable to diameter/gap) was designed intentionally to enhance the sensitivity to the change of surface property. In other words, the capacitor of shorter height is more sensitive to the changes of the dielectric constant ε and the surface morphology of the sensor during the course of corrosion, because of greater specific surface area. Although Equation (13) does not consider the effect of fringe field, which lacks of a reliable equation to quantify, it does describe the fundamental relationship between capacitance C and the dielectric constant ε_1 and the average thickness of the rust layer δ, which is verified by the experimental results from this study.

During corrosion monitoring of a structure, the empirical equations can be developed based on laboratorial tests in an environmental chamber through multiple regressions, *i.e.*, the differences in resistance and capacitance readings between the corrosion and the reference sensors as a function of the parameters including the extent of corrosion (iron loss), temperature and moisture level. The equations are used to interpret the resistance and capacitance data from the site along with the site's temperature and moisture information. As a result, the extent of corrosion can be determined by plugging the site's temperature and moisture data in the equation and then solving the extent of corrosion numerically. In addition, depending on the specific requirement of a monitoring site, a feature of periodic sleep and wakeup time can be adopted to save energy and cost.

In order to mitigate potential fouling problem of the sensors on site, paired corrosion and reference sensors are installed side-by-side with the identical coating/passivation in order to minimize the uncertainty brought by location-dependent fouling issue caused by such as particles and debris. Special cares are taken to ensure the orientation of the sensors is not prone to accumulate dust; while rainfall and snow melting can help clean the sensors. After normalizing the capacitance of both of the corrosion and the reference sensors, the systematic differences in resistance and capacitance readings between the corrosion and the reference sensors are expected to reflect the extent of corrosion. In addition, sufficient paired corrosion and reference sensors are attached to the structure being monitored. Statistical analysis is an important part of the corrosion monitoring network. The averaged differences in resistance and capacitance readings between the corrosion and the reference sensors can further minimize the uncertainty caused by fouling problem.

EXPERIMENTAL PROCEDURES

Cylindrical Capacitor

ASTM A36 steel was used in this study. It contains at least 99.05% of Fe, and max 0.26% C, max 0.04% P, max 0.05% S, max 0.40% Si, and max 0.20% Cu (by wt) [33]. As shown in Figure 5, a cylindrical capacitor was created with the inner cylinder made from A36 carbon steel rod and the outer ring made from 316 stainless steel. Each of them had a height of 0.64 cm. The inner cylinder had a diameter of 1.27 cm while the outer ring had an outer diameter of 2.64 and inner diameter of 2.22 cm. The A36 carbon steel was polished by a 3M® 80 grit then a 3M® 600 grit sandpaper. Two wires were attached to the capacitor, one was soldered to the base of the center A36 steel rod and the other was welded to the base of the outer ring of the 316 stainless steel. A bridge made of a glass substrate epoxy resin insulator from the circuit board (BM-FR4-1SS2, T-Tech Inc., Norcross, GA, USA) was adhered on the bottom of both the inner cylinder and the ring through a waterproof epoxy (15206 Anchor-Tite, Super Glue Corporation, Rancho Cucamonga, CA, USA) to fix their relative positions. Finally, the waterproof epoxy was used to cover the connections of these wires as well as the base of the sensor to prevent corrosion of the connections and the wires, as well as to insulate the sensor from the infrastructure to be monitored. The uncovered A36 steel had a surface area of about 2.66 cm^2, which was subject to corrosion.

Similarly, a reference sensor was made following the same procedures and dimensions except replacing the A36 carbon steel rod with a 316 stainless steel rod of the same dimensions, and then being welded to connect to the circuit-board bridge. The connections including welding points of the reference sensor were coated with the waterproof epoxy to prevent corrosion. The reference sensor was served to draw baseline information by addressing environmental conditions such as the temperature and the moisture level of air other than corrosion.

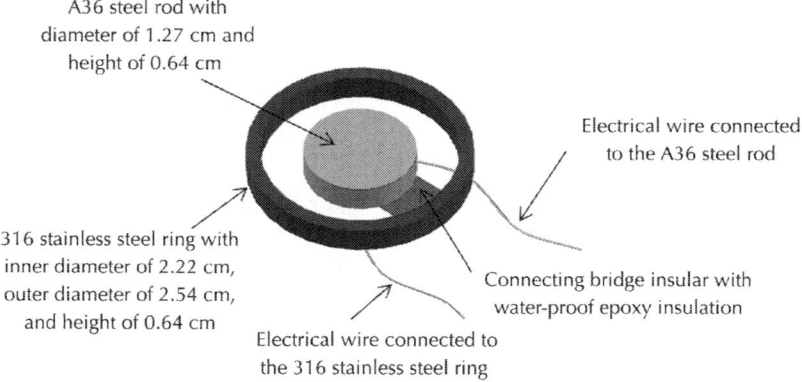

A36 steel rod with diameter of 1.27 cm and height of 0.64 cm

Electrical wire connected to the A36 steel rod

316 stainless steel ring with inner diameter of 2.22 cm, outer diameter of 2.54 cm, and height of 0.64 cm

Connecting bridge insular with water-proof epoxy insulation

Electrical wire connected to the 316 stainless steel ring

Figure 5: Diagram of the prototype corrosion sensor made of a cylindrical capacitor used for corrosion monitoring.

Corrosion Test

A 500-mL 0.2 M sodium chloride solution at 20 ± 1 °C was used for corrosion testing. An air pump (Aqua Culture®, China) with a flow rate of ~1.2 L/min was continuously bubbling air through a diffuser to the solution to provide oxygen for the corrosion process (Figure 6). The air diffuser was porous sandstone. The dissolved oxygen level of the NaCl solution was maintained at around 8.8 mg/L. The corrosion sensor was submerged in the sodium chloride solution above the air diffuser. Every day during the course of the test, the sensor was removed from the solution, rinsed with DI water, dried at room temperature for 2–3 h, and then tested with an automatic RCL meter (PM6303A, Fluke Corporation, Everett, WA, USA). At each measurement, multiple readings from the RCL meter with time were recorded until stable readings obtained, indicating the senor was dried at equilibrium with air moisture. As a result, the sensor experienced periodically wet/dry cycles twice a day for total 11 days. A new 0.2 M sodium chloride solution was freshly made daily. The accumulated corrosion time of the senor in the sodium chloride solution was 225.5 h. As a control test, a 316 stainless steel ring and the reference sensor was soaked in an aerated 500-mL 0.2 M sodium chloride solution separately to investigate the degree of corrosion of the 316 stainless steel and the reference sensor, respectively.

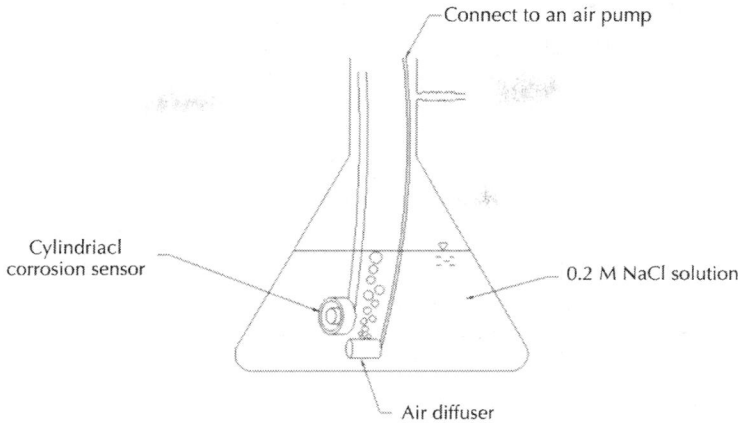

Connect to an air pump

Cylindriacl
corrosion sensor

0.2 M NaCl solution

Air diffuser

Figure 6: The corrosion testing device. The prototype cylindrical sensor was submerged in an aerated 0.2 M NaCl solution with an air diffuser underneath.

Sensor Measurements

The RCL meter was used to measure the resistance and the capacitance of the prototype corrosion sensors or the reference sensors. The readings were obtained in parallel mode, which was one of the built-in functions of the meter. The meter used an AC power with frequency of 1 kHz and a voltage of 1.9 V for the measurements to minimize electrolysis and electrode polarization problems brought by a DC power. With an increase in AC frequency, the interval between successive anodic and cathodic half-cycles becomes progressively shorter [34]. Thus the electrolytic reactions do not have enough time to complete, and it even increases the potential to reverse the reactions of the immediate prior cycle. When the power frequency is high enough (e.g., 1 kHz [35]), no electrolysis of AC was observed, because all current would pass via the double layer of the electrodes [34,36]. The RCL meter took five measurements: a reference reading, a voltage reading at the phase angle of 0° and 90°, and a current reading at the phase angle of 0° and 90°, it then calculated the values for resistance and capacitance (which could be in either series and parallel) based on this model. It was decided to use the parallel mode of this meter as the focus of these measurements is on a capacitor, so all the measurements in this study were found using the parallel mode. In real time monitoring of an infrastructure, instead of using a RCL meter, the corrosion sensor data acquisition algorithm, coded in VbScript, running under a Windows XP embedded computer, produced one data file for each sensor at 30 min interval (see Supporting Information).

In addition, the mass of the dried sensor was examined by a digital balance (ESA-3000, Brecknell, Fairmont, MN, USA) with a precision of 0.05 g. The initial weight of the prototype corrosion sensor was 25.20 g. For the prototype corrosion sensor, the initial resistance (R_0) was $48.63 \times 10^6 \ \Omega$; the initial capacitance (C_0) was 7.8 pF before corrosion test. For the stainless steel reference sensor, the initial resistance was $37.66 \times 10^6 \ \Omega$; the initial capacitance (C_0) was 24.9 pF. The difference of the initial readings between the corrosion and the reference sensor is likely due to welding connection to 316 stainless steel rod (rather than soldering to A36 carbon steel), because soldering is not feasible for stainless steel.

Sample Analyses

After the corrosion sensor was removed from the NaCl solution, the daily solution samples were acidified with ACS grade nitric acid from Mallinckrodt to 10% (v/v) to dissolve the iron rust in the solution. AAS (AAnalyst 200, PerkinElmer, Waltham, MA, USA) was utilized to measure the total dissolved iron in the solution.

The crystalline substances on both of the corroded and uncorroded A36 steel surface were examined by XRD with CuKα radiation (APD3520, Philips, Amsterdam, The Netherlands). For the uncorroded sample, a piece of the A36 steel of 1.27 cm diameter and 0.2 cm thickness was polished by a 3M® 80 grit then a 3M® 600 grit sandpaper. The steel sample was cleaned by blowing thoroughly with compressed nitrogen. For the corroded sample, the same polishing procedures were followed as the uncorroded sample, then the A36 steel sample was submerged in an aerated 500-mL 0.2 M sodium chloride solution for 225.5 h. The corroded steel sample was rinsed with DI water and dried in air for XRD examination.

CONCLUSIONS

Automatic detection of the early-stage corrosion is highly important to find the potential problem and apply corrosion control techniques timely for safety and integrity concerns. This study explored an innovative cylindrical corrosion sensor made of A36 carbon steel (representing the material of a structure or a system to be monitored for corrosion) and a 316 stainless steel ring (representing an inert material of low corrosion potential). A capacitor was formed with both conductors separated by air. The sensor is more sensitive than the conventional corrosion sensors based on the bulk electrical-resistance method. After corrosion in an aerated 0.2 M NaCl solution for 225.5 h, the cylindrical corrosion sensor has shown a systematic decrease in the normalized electrical resistance (R/R_0) from 1.0 to 0.74. Meanwhile, the normalized capacitance (C/C_0) of the sensor increased from 1.0 to 1.46. However, the weight change of the sensor was within 0.2% (or 187.7 g/m^2), an indication of the early-stage corrosion. In the same time, the reference senor, which was not subject to corrosion apparently, showed a stable normalized reading around 1.0. XRD result shows that the rust contained lepidocrocite and magnetite. By attaching the paired corrosion and reference sensors with the identical passivation/coating to a steel structure in air, the extent of corrosion of the structure can be directly reflected by the electrical resistance decrease and/or capacitance increase of the sensor with time during the early-stage corrosion.

ACKNOWLEDGMENTS

The research project was funded by the US Army Engineer Research and Development Centre, Construction Engineering Research Laboratory through the contract of W9132T-10-2-0056. The authors also thank Robert Tilbury and Mengwei Li for their help for this project. XRD was performed in the Argast Family Instrumentation and Analysis Lab at the Indiana University-Purdue University Fort Wayne.

AUTHOR CONTRIBUTIONS

In this paper, Dong Chen and Max Yen formulated the research ideas and the test plan. Paul Lin designed and built the data collection, the processing and the storage system. Steve Groff carried out the corrosion experiments in lab and consolidated the data. Richard Lampo, Michael McInerney and Jeffrey Ryan supervised the research, provided the technical guidance and prepared the experimental materials.

REFERENCES

1.   Uhlig, H.H.; Revie, R.W. *Corrosion and Corrosion Control: An Introduction to Corrosion Science and Engineering*; John Wiley & Sons: New York, NY, USA, 1985.

2.   Borgard, B.; Warren, C.; Somayaji, S.; Heidersbach, R. In *Corrosion Rates of Steel in Concrete*; ASTM STP 1065. Berke, N.S., Chaker, V., Whiting, D., Eds.; American Society for Testing and Materials: Philadelphia, PA, USA, 1990; pp. 174–188.

3.   Fontana, M.G.; Greene, N.D. *Corrosion Engineering*; McGraw-Hill: New York, NY, USA, 1967. Chapter 1.

4.   Soares, C.G.; Garbatov, Y.; Zayed, A. Effect of environmental factors on steel plate corrosion under marine immersion conditions.*Corros. Eng. Sci. Technol.* 2011, *46*, 524–541.

5.   Montemor, M.F.; Simões, A.M.P.; Ferreira, M.G.S. Chloride-induced corrosion on reinforcing steel: From the fundamentals to the monitoring techniques. *Cem. Concr. Compos.* 2003, *25*, 491–502.

6.   Haleem, S.M.A.E.; Wanees, S.A.E.; Aal, E.E.A.E.; Diab, A. Environmental factors affecting the corrosion behavior of reinforcing steel II. Role of some anions in the initiation and inhibition of pitting corrosion of steel in $Ca(OH)_2$ solutions. *Corros. Sci.* 2010,*52*, 292–302.

7.   Morris, W.; Vico, A.; Vázquez, M. Chloride induced corrosion of reinforcing steel evaluated by concrete resistivity measurements. *Electrochim. Acta* 2004, *49*, 4447–4453.

8. Li, S.Y.; Kim, Y.G.; Jung, S.; Song, H.S.; Lee, S.M. Application of steel thin film electrical, resistance sensor for in situ corrosion monitoring. *Sens. Actuators B* 2007, *120*, 368–377.

9. Zivica, V. Utilisation of electrical resistance method for the evaluation of the state of steel reinforcement in concrete and the rate of its corrosion. *Constr. Build. Mater.* 2000, *14*, 351–358.

10. Reading, M.S.; Denzine, A.F. A critical comparison of corrosion monitoring techniques used in industrial applications. In *Industrial Corrosion and Corrosion Control Technology*; Shalaby, H.M., Al-Hashem, A., Lowther, M., Al-Besharah, J., Eds.; Kuwait Institute for Scientific Research: Kuwait City, Kuwait, 1996; pp. 511–519.

11. Popova, A.; Veleva, S.; Raicheva, S. Kinetic approach to mild steel corrosion. *React. Kinet. Catal. Lett.* 2005, *85*, 99–105.

12. Fontana, M.G. *Corrosion Engineering*, 3rd ed.; McGraw Hill: New York, NY, USA, 1987.

13. Sen, R.; Liby, L.; Mullins, G. Strengthening steel bridge sections using CFRP laminates. *Compos. Part B Eng.* 2001, *32*, 309–322.

14. Tavakkolizadeh, M.; Saadatmanesh, H. Fatigue strength of steel girders strengthened with carbon fiber reinforced polymer patch. *J. Struct. Eng. ASCE* 2003, *129*, 186–196.

15. Evans, U.R.; Taylor, C.A.J. Mechanism of atmospheric rusting. *Corros. Sci.* 1972, *12*, 227–246.

16. Haupt, S.; Strehblow, H.H. Corrosion, layer formation, and oxide reduction of passive iron in alkaline solution: A combined electrochemical and surface analytical study. *Langmuir* 1987, *3*, 873–885.

17. Abd El-Maksoud, S.A. The effect of organic compounds on the electrochemical behavior of steel in acidic media. *Int. J. Electrochem. Sci.* 2008, *3*, 528–555.

18. Broomfield, J.P. *Corrosion of Steel in Concrete: Understanding, Investigation and Repair*; Taylor & Francis: London, UK, 2007. Chapter 2.

19. Roberts, W.L.; Campbell, T.J.; Rapp, G.R., Jr. *Encyclopedia of Minerals*, 2nd ed.; Chapman & Hall: New York, NY, USA, 1990.

20. In *Mineral Powder Diffraction File: Data Book*; JCPDS-International Centre for Diffraction Data. American Society for Testing and Materials: Swarthmore, PA, USA, 1986.

21.   Suzuki, I.; Masuko, N.; Hisamatsu, Y. Electrochemical properties of iron rust. *Corros. Sci.* 1979, *19*, 521–535.

22.   González, J.A.; Miranda, J.M.; Otero, E.; Feliu, S. Effect of electrochemically reactive rust layers on the corrosion of steel in a Ca(OH)$_2$ solution. *Corros. Sci.* 2007, *49*, 436–448.

23.   Nishimura, N.; Katayama, H.; Noda, K.; Kodama, T. Electrochemical behavior of rust formed on carbon steel in a wet/dry environment containing chloride ions. *Corrosion* 2000, *56*, 935–941.

24.   Yuan, J.; Wu, X.; Wang, W.; Zhu, S.; Wang, F. The effect of surface finish on the scaling behavior of stainless steel in steam and supercritical water. *Oxid. Met.* 2013, *79*, 541–551.

25.   Varshney, D.; Yogi, A. Structural and electrical conductivity of Mn doped hematite (α-Fe$_2$O$_3$) phase. *J. Mol. Struct.* 2011, *995*, 157–162.

26.   Lair, V.; Antony, H.; Legrand, L.; Chaussé, A. Electrochemical reduction of ferric corrosion products and evaluation of galvanic coupling with iron. *Corros. Sci.* 2006, *48*, 2050–2063.

27.   Ltai, R.; Shibuya, M.; Matsumura, T.; Ishi, G. Electrical resistivity of magnetite anodes. *J. Electrochem. Soc.* 1971, *118*, 1709–1711.

28.   Akl, A.A. Microstructure and electrical properties of iron oxide thin films deposited by spray pyrolysis. *Appl. Surf. Sci.* 2004, *221*, 319–329.

29.   Glotch, T.D.; Rossman, G.R. Mid-infrared reflectance spectra and optical constants of six iron oxide/oxyhydroxide phases. *Icarus*2009, *204*, 663–671.

30.   Filius, J.D.; Lumsdon, D.G.; Meeussen, J.C.L.; Hiemstra1, T.; van Riemsdijk, W.H. Adsorption of fulvic acid on goethite.*Geochim. Cosmochim. Acta* 2000, *64*, 51–60.

31.   Haynes, W.M. *CRC Handbook of Chemistry and Physics,*, 91st ed.; CRC Press: Boca Raton, FL, USA, 2010.

32.   Basu, B.N. *Electromagnetic Theory and Applications in Beam-Wave Electronics*; World Scientific: Singapore, Singapore, 1996.

33.   In *Standard Specification for Carbon Structural Steel*; ASTM Standard A36/A36M-08. ASTM International: West Conshohocken, PA, USA, 2008.

34. Park, J.C.; Lee, M.S.; Han, D.W.; Lee, D.H.; Park, B.J.; Lee, I.S.; Uzawa, M.; Aihara, M.; Takatori, K. Inactivation of Vibrio parahaemolyticus in effluent seawater by alternating-current treatment. *Appl. Environ. Microbiol.* 2004, *70*, 1833–1835.

35. Kortum, G. *Treatise on Electrochemistry*; Elsevier: London, UK, 1965.

36. Fernandes, S.Z.; Mehendale, S.G.; Verkatachalam, S. Influence of frequency of alternating current on the electrochemical dissolution of mild steel and nickel. *J. Appl. Electrochem.* 1980, *10*, 649–654.

Citations

CHAPTER 1

David W. Hoeppner and Carlos A. Arriscorreta, "Exfoliation Corrosion and Pitting Corrosion and Their Role in Fatigue Predictive Modeling: State-of-the-Art Review," International Journal of Aerospace Engineering, vol. 2012, Article ID 191879, 29 pages, 2012. doi:10.1155/2012/191879.

CHAPTER 2

E. Mohseni, E. Zalnezhad, Ahmed A. D. Sarhan, and A. R. Bushroa, "A Study on Surface Modification of Al7075-T6 Alloy against Fretting Fatigue Phenomenon," Advances in Materials Science and Engineering, vol. 2014, Article ID 474723, 17 pages, 2014. doi:10.1155/2014/474723.

CHAPTER 3

Iwona B. Beech and Christine C. Gaylarde, Recent Advances in the Study of Biocorrosion - an Overview, http://dx.doi.org/10.1590/S0001-37141999000300001.

CHAPTER 4

C. E. B. Marino and L. H. Mascaro, "Electrochemical Tests to Evaluate the Stability of the Anodic Films on Dental Implants," International Journal of Electrochemistry, vol. 2011, Article ID 574502, 7 pages, 2011. doi:10.4061/2011/574502.

CHAPTER 5

Kai Wang Chan and Sie Chin Tjong, Effect of Secondary Phase Precipitation on the Corrosion Behavior of Duplex Stainless Steels, doi:10.3390/ma7075268.

CHAPTER 6

Han-Seung Lee, Hwa-Sung Ryu, Won-Jun Park, and Mohamed A. Ismail, Comparative Study on Corrosion Protection of Reinforcing Steel by Using Amino Alcohol and Lithium Nitrite Inhibitors, doi:10.3390/ma8010251.

CHAPTER 7

Dong Chen, Max Yen, Paul Lin, Steve Groff, Richard Lampo, Michael McInerney, and Jeffrey Ryan, A Corrosion Sensor for Monitoring the Early-Stage Environmental Corrosion of A36 Carbon Steel, doi:10.3390/ma7085746.

Index